Advance Praise for

HEADSTRONG

"A woman revolutionized heart surgery. A woman created the standard test given to all newborns to determine their health. A woman was responsible for some of the earliest treatments of previously terminal cancers. We shouldn't need to be reminded of their names, but we do. With a deft touch, Rachel Swaby has assembled an inspiring collection of some of the central figures in twentieth-century science. *Headstrong* is an eye-opening, much-needed exploration of the names history would do well to remember, and Swaby is a masterful guide through their stories."

—Maria Konnikova, *New Yorker* writer and *New York Times* bestselling author of *Mastermind*

"*Headstrong* is a true gem. So many amazing women have had an incredible impact on STEM fields, and this book gives clear, concise, easy-to-digest histories of fifty-two of them—there's no longer an excuse for not being familiar with our math and science heroines. Thank you, Rachel!"

—Danica McKellar, actress and *New York Times* bestselling author of *Math Doesn't Suck*

"Rachel Swaby's fine, smart look at women in science is a much-needed corrective to the record—a deftly balanced field guide to the overlooked (Hilde Mangold), the marginalized (Rosalind Franklin), the unexpected (Hedy Lamarr), the pioneering (Ada Lovelace), and the still-controversial (Rachel Carson). Swaby reminds us that science, like the rest of life, is a team sport played by both genders."

—William Souder, author of *On a Farther Shore* and *Under a Wild Sky*

HEADSTRONG

HEADSTRONG

52
WOMEN
WHO CHANGED
SCIENCE—
AND THE
WORLD

RACHEL SWABY

B\D\W\Y
Broadway Books
New York

Published in the United States by Broadway Books,
an imprint of the Crown Publishing Group, a division
of Penguin Random House LLC, New York.
www.crownpublishing.com

Permission acknowledgments can be found on pp. 271–273.

Library of Congress Cataloging-in-Publication Data
Swaby, Rachel.
Headstrong : 52 women who changed science—and
the world / Rachel Swaby.—First edition.
pages cm
Includes bibliographical references and index.
1. Women scientists—Biography. 2. Women astronomers—Biography.
3. Women physicians—Biography. 4. Women biologists—Biography.
5. Women physicists—Biography. 6. Women mathematicians—Biography.
I. Title.
Q130.S93 2015
509.2'52—dc23 2014041421

ISBN 978-0-553-44679-1
eBook ISBN 978-0-553-44680-7

Printed in the United States of America

Cover design by Na Kim
Cover photographs: Clockwise from top left: (Sally Ride) U.S. National Archives and Records Administration/Wikimedia Commons/Public Domain; (Chien-Shiung Wu) Smithsonian Institution Archives. Image SIA2010-1507; (Hedy Lamaar) Wikimedia Commons/Public Domain; (Yvonne Brill) Walter P. Reuther Library, Archives of Labor and Urban Affairs, Wayne State University; (Rita Levi-Montalcini) Mondadori Portfolio; (Alice Ball) Wikimedia Commons/Public Domain; (Ellen Richards) Wikimedia Commons/Public Domain; (Annie Cannon) Library of Congress [Reproduction number LC-USZ62-115881]; (Dorothy Hodgkin) Peter Lofts Photography/National Portrait Gallery, London; (Florence Nightingale) Library of Congress Prints and Photographs Division [Reproduction number LC-DIG-ppmsca-037769]; (Maria Mitchell) *Sweeper in the Sky: The Life of Maria Mitchell, First Woman Astronomer in America* by Helen Wright, 1914–1997. New York: The Macmillan Company, 1949. Wikimedia Commons/Public Domain.

10

First Edition

For Tim

CONTENTS

PHYSICS

EARTH AND STARS

MATH AND TECHNOLOGY

INVENTION

INTRODUCTION

THIS BOOK ABOUT SCIENTISTS BEGAN WITH BEEF STROGANOFF. According to the *New York Times,* Yvonne Brill made a mean one. In an obituary published in March 2013, Brill was honored with the title of "world's best mom" because she "followed her husband from job to job and took eight years off from work to raise three children." Only after a loud, public outcry did the *Times* amend the article so it would begin with the contribution that earned Brill a featured spot in the paper of record in the first place: "She was a brilliant rocket scientist." Oh right. *That.*

The error—stroganoff before science; domesticity before personal achievement—is so cringe-worthy because it's a common one. In 1964, when Dorothy Crowfoot Hodgkin won the greatest award that chemistry has to offer, a newspaper declared "Nobel Prize for British Wife," as if she had stumbled upon the complex structures of biochemical substances while matching her husband's socks. We simply don't speak of men in science this way. Their marital status isn't considered necessary context in a biochemical breakthrough. Employment as an important aerospace engineer is not the big surprise hiding behind a warm plate of noodles. For men, scientific accomplishments are accepted as something naturally within their grasp.

In 1899, the inventor and physicist Hertha Ayrton put on a demonstration showing her latest breakthrough in calming the temperament of the arc light, long notorious for hissing and flickering. When the newspaper reported on the presentation, it treated Ayrton like some kind of circus performer: "What astonished the lady visitors . . . was to find one of their own sex in charge of the most dangerous-looking of all the exhibits. Mrs.

Ayrton was not a bit afraid." Annoyed by this and many other similar perspectives, Ayrton called out a persistent problem with the way she and her contemporaries like Marie Curie were treated: "The idea of 'women and science' is entirely irrelevant. Either a woman is a good scientist or she is not; in any case she should be given opportunities, and her work should be studied from the scientific, not the sex, point of view."

Even today, it's important we hear those words again. We need not only fairer coverage of women in science, but more of it.

Access to role models really matters for girls coming up in the STEM fields. Sally Ride turned her father into an advocate for the cause. After coming across an advertisement featuring a boy daydreaming about the day he would go up into space, Ride's father wrote a strongly worded letter to the advertiser pointing out an inherent bias in educating children that should be corrected. "As a parent of the first US woman astronaut, I know first hand that girls also aspire to math and science and we should encourage her to 'get America's future off the ground.'" In a *New York Times Magazine* article, Eileen Pollack, one of the first two women to earn a bachelor's degree in physics at Yale, noted the large poster of famous mathematicians hanging in her alma mater's math department lobby, which—even at the time of the article, in 2013—didn't include a single woman. She had opted not to continue on in science. In early 2014, a seven-year-old named Charlotte wrote an open letter to Lego. "I went to a store and saw Legos in two sections, the pink [girls] and the blue [boys]. All the girls did was sit at home, go to the beach, and shop and they had no jobs but the boys went on adventures, worked, saved people, and had jobs, even swam with sharks. I want you to make more Lego girl people and let them go on adventures and have fun ok!?!"

As girls in science look around for role models, they shouldn't have to dig around to find them. By treating women in science like scientists instead of anomalies or wives who moonlight in the lab as well as correcting the cues given to girls at a young age about what they're good at and what they're supposed to like, we can accelerate the growth of a new generation of chemists, archeologists, and cardiologists while also revealing a hidden history of the world.

By her own standards, Hertha Ayrton was a good scientist. So was the detail-oriented seismologist Inge Lehmann, and the firecracker neuroembryologist Rita Levi-Montalcini, too. The scientists in this book aren't included because they were women practicing science or math in a time when few women did—although by that criterion, many would fit. They're included because they discovered Earth's inner core, revealed radioactive elements, dusted off a complete dinosaur skeleton, or launched a new field of scientific inquiry. Their ideas, discoveries, and insights made earth-shaking changes to the way we see the world (and that goes for the seismologist, too).

Accomplishments alone could have warranted inclusion in a different kind of book, but to be here, narrative—a secret bedroom lab, an ocean-floor expedition, or a stolen photograph that helped solve the structure of DNA—needed to be the twin pillar of achievement. Bullet points of a dazzling career weren't enough.

To make sure each subject's lasting influence is clear, the book includes only scientists whose life's work has already been completed. Omitting the living was particularly painful since it meant filtering out so many extraordinary scientists and achievements. Furthermore, opportunities for white women in STEM fields opened up before they did for women of color. Even five years from now, a book with much greater diversity would emerge from the same criteria.

Because Marie Curie is who we talk about when we talk

about women in science, I've chosen not to include her. She's the overwhelming favorite for almost every occasion: the token woman in a deck of cards featuring famous scientists, the one most likely to pop up in casual conversation, and the scientist to which all other women in science are compared. A two-time Nobel Prize winner, the director of Paris's hugely influential Radium Institute, and the scientist who first drew widespread public attention to this little prize called the Nobel, Curie certainly deserves her place in history and in our zeitgeist. For Chien-Shiung Wu, Marguerite Perey, and even her own daughter, Irène Joliot-Curie, Marie Curie was an inspiration. My hope is that the stories in this book will provide readers of every age a new set of scientists, mathematicians, and engineers to admire.

So instead of calling every standout woman in science the Marie Curie of her field, the next time someone really lives for their work, let's call them the Barbara McClintock of their specialty. If a scientist charts new territory, let's refer to them as the Annie Jump Cannon of their particular exploration. If a researcher puts herself in physical danger for an experiment, let's say she's like any number of the scientists here who worked with radioactivity or mustard gas.

There are fifty-two profiles in this book. Read one a week, and in a year you'll know whose research jump-started the Environmental Protection Agency, who discovered wrinkle-free cotton, and even whose ingenious score has now saved generations of struggling newborns. So little coverage has been dedicated to these scientists elsewhere that, in going through these profiles, I hope you'll feel like you've gained a breadth of knowledge that rivals that of Salome Waelsch.

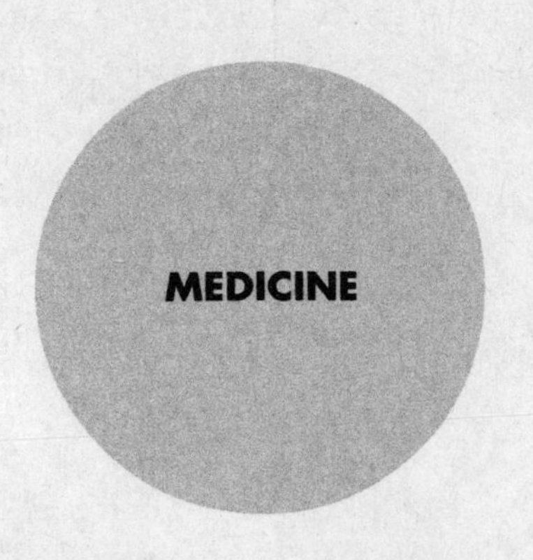
MEDICINE

MARY PUTNAM JACOBI

1842–1906

MEDICINE • AMERICAN

A WARNING FROM EDWARD CLARKE, MD, PROFESSOR AT HARvard: "There have been instances, and I have seen such, of females . . . graduated from school or college excellent scholars, but with undeveloped ovaries. Later they married, and were sterile." He goes on to explain how reproductive organs fail to thrive. "The system never does two things well at the same time. The muscles [note: muscles = menstruation] and the brain cannot *functionate* in their best way at the same moment." These passages are from Clarke's book, *Sex in Education; or, A Fair Chance for Girls*, published in 1873. The gist: Exerting oneself while on the rag is dangerous. Therefore educating women is dangerous. For a woman's own safety, she should not pursue higher education. The womb is at stake.

Today, it's easy to write off Clarke's thesis as one doctor's nutty ramblings. His descriptions of students—"crowds of pale, bloodless female faces, that suggest consumption, scrofula, anemia, and neuralgia," as a result of "our present system of educating girls"—sounds more like *The Walking Dead* than students at a university campus. But when *A Fair Chance for Girls* was published, administrators and faculty opposed to women in education hoisted up the book as a confirmation of their views, couched in an argument about safety.

Mary Putnam Jacobi thought the whole thing was hogwash. Jacobi, an American, was the first woman admitted to France's École de Médecine. It took a bit of wrangling, but once she was in, Jacobi found her medical training thrilling. Certainly there were people who doubted her ability to succeed—even her mother did some hand-wringing over her schooling—but Jacobi proceeded with ease and humor. In 1867, she wrote home to assure her mother, "I really am only enjoying myself . . . the hospitals present so much that is stimulating, (and do not be shocked if I add amusing) that I am never conscious of the slightest head strain."

To battle Clarke's assertions, Jacobi could have presented her personal experience as a counterargument. Her education at the École de Médecine took place *after* she'd already received an MD in the United States. Medical school made Jacobi neither ill nor infertile. But bringing forward an autobiographical account when evidence was within reach was like feeling for your own heartbeat when it could be measured with a stethoscope.

Jacobi challenged Clarke's thinly veiled justification for discrimination with 232 pages of hard numbers, charts, and analysis. She gathered survey results covering a woman's monthly pain, cycle length, daily exercise, and education along with physiological indicators like pulse, rectal temperature, and ounces of urine. To really bring her argument home, Jacobi had test subjects undergo muscle strength tests before, during, and after menstruation. The paper was almost painfully evenhanded. Her scientific method–supported mic drop: "There is nothing in the nature of menstruation to imply the necessity, or even the desirability, of rest." If women suffered from consumption, scrofula, anemia, and neuralgia, it wasn't, as Clarke claimed, because they studied too hard.

Her report—sweeter for its evidence than its tone—won the Boylston Prize at Harvard University just three years after Clarke, a professor at the same school, published *A Fair Chance*. The Clarke versus Jacobi scholarly disagreement wasn't just academic quibbling, a biased doctor against a rigorous one. In the argument over who was allowed university admission, to have science on your side was hugely important. After Clarke's paper fortified university walls, Jacobi systematically dismantled the barrier. Her paper was greatly influential in helping women gain opportunities in higher education—especially in the sciences.

Jacobi had wanted to be a doctor since childhood. "I began my medical studies when I was about nine years old," she remembered. "I found a big dead rat and the thought occurred to me that if I had the courage, I could cut that rat open and could find his heart which I greatly longed to see . . . my courage failed me." Although she tabled the exploratory surgery until she was trained to do it, her interest in the body never waned. In the meantime, Jacobi wrote. Growing up in a family of well-known book publishers, she dabbled in her family's business, placing stories in *The Atlantic Monthly* starting at age fifteen and later in the *New-York Evening Post*.

Jacobi's father wasn't thrilled to hear she'd decided to attend medical school. In response, he dangled the amount of her university tuition before her, a carrot that would be hers should she decide against higher education. Jacobi declined his offer, leaving for the Woman's Medical College of Pennsylvania, in the early 1860s, before continuing on to Paris for a second round of schooling. When her mother wrote requesting an update, Jacobi replied, "I think you are rather naive to ask me if 'I meet many educated French ladies who are physicians.' Such a thing was never heard of."

In Paris, an American was considered enough of a curiosity that after months of lobbying, Jacobi was able to claim the first spot at the École de Médecine ever awarded to a woman. There were a few stipulations attached to her attendance. She had to enter lectures through a door not used by other students and sit near the professor. Jacobi joked that hers would be the first petticoat the school had seen since its founding. However strange the circumstances, Jacobi found assimilation easy. She wrote, "I . . . feel as much at home as if I had been there all my life."

Upon returning to the United States after five years in Paris, Jacobi began lecturing at Women's Medical College of the New York Infirmary for Women and Children, practicing medicine and carving out more opportunities for women in the field concurrently. Jacobi helped found the Women's Medical Association of New York City in 1872, opened New York Infirmary's children's ward, and became the Academy of Medicine's first female member. When she was diagnosed with a brain tumor, Jacobi documented the symptoms just as thoroughly and objectively as if she were responding to Clarke's ridiculous claims. She titled the result, "Description of the Early Symptoms of the Meningeal Tumor Compressing the Cerebellum. From Which the Writer Died. Written by Herself." Jacobi did always like to get in the last word.

ANNA WESSELS WILLIAMS

1863–1954

BACTERIOLOGY • AMERICAN

ALTHOUGH ANNA WESSELS WILLIAMS BELIEVED COLLABORATION was essential, she seized every moment of solitude. In her free time, she went up in stunt planes, teetering between the stomach-dropping danger of pre–World War I flight and the sublime feeling of gliding where so few others could. On the ground, she piled up speeding tickets, the allure of zipping around those in her way apparently too tempting to resist. Williams was alone, this time in the New York City Department of Health's diagnostic laboratory, when she made one of the lab's greatest discoveries. In 1894, she isolated a strain of the infectious disease diphtheria. That strain became crucial in developing higher yields of an antitoxin needed for fighting the infection.

Diphtheria is under control now, but when Williams was working on the problem, it had reached "near-epidemic levels." The disease hitched a ride on spittle transferred from one person to the next during a cough or during conversation. At first diphtheria just caused fever or chills, but when it really settled in, it could wreak havoc on the heart and nervous system. Children were dying, and those living in poverty were disproportionately at risk.

An antitoxin for diphtheria had been discovered just four years earlier in 1890 by Emil von Behring. It was a major accomplishment, a breakthrough that would earn von Behring the Nobel Prize in Medicine in 1901. Finding a method of infection therapy is one thing, but deploying it globally is entirely another. His antitoxin needed a toxin to activate it, and in the intervening years, scientists were stuck with a low yield of the starter. There wasn't enough of the serum to go around. Meanwhile, the disease continued to leap from person to person, killing thousands of children per year.

Under the guidance of William H. Park at the Department of Health's diagnostic laboratory, Anna Williams got to work on finding a strain of the bacteria that could bring forth a powerful toxin to activate the antitoxin, and in high enough volume to produce it at scale. The breakthrough came while Park was away on vacation. Williams isolated a strain of bacteria that could generate a toxin 500 times more potent than what was previously available.

The strain was named Park-Williams No. 8. Gracious about her boss's inclusion, she said she was "happy to have the honor of having my name thus associated with Dr. Park." Williams recognized the necessity of collaboration in research. After all, Williams's own experiments were bolstered by von Behring's initial breakthrough. But over time, Park-Williams No. 8 proved too many syllables for the people who worked with it; informally, it became known as Park 8. Just like that, Williams's pioneering work was clipped from view.

Name recognition wasn't why Williams got into science; she wasn't concerned with how many strains of bacteria her name was slapped onto. Her sense of purpose originated from addressing a medical need. Where real-world good was concerned, Park 8 succeeded spectacularly. Since the new strain increased the antitoxin's availability and slashed cost, Williams was instrumental in slowing down the spread of disease. Within a year of her discovery, the diphtheria antitoxin went into mass production. To address the staggering demand, mass quantities of the preparation were shipped to physicians in the United States and England without charge.

Williams's decision to go into medical science originated from a time when she witnessed one bad event spin out of control without the knowledge or training to intervene. For Williams, the stillbirth of her sister's baby in 1887 and the near death of her sister during childbirth would have at least been partially avoidable had the attending physician been more thoroughly trained. The horrific incident gave Williams resolve. She would battle such medical ignorance with her own education.

Williams quit her job as a schoolteacher almost immediately after the incident in order to enroll in classes at the Women's Medical College of the New York Infirmary. She found her schoolwork thrilling: "I was starting on a way that had been practically untrod before by a woman. My belief at the time in human individuality, regardless of sex, race, religion or any factor other than ability was at its strongest. I believed, therefore, that females should have equal opportunities with males to develop their powers to the utmost." By 1891, she'd earned her MD.

At the New York City Department of Health, opportunities to tackle nasty diseases appeared almost immediately. That diphtheria breakthrough? It happened in 1894, within her first year at the organization, when she was still just a volunteer. She was added to the payroll the following year and given a title: assistant bacteriologist.

Creative and fearless, Williams took a sabbatical in Paris in 1896 to research scarlet fever at the Pasteur Institute. There she was met with a culture of deep secrecy, where discussion of time-sensitive research was strictly off-limits and research tools like cadavers were not to be shared. She hoped to do for scarlet fever what she'd done for diphtheria, but the research was a bust.

The trip was redeemed by rabies, or rather by the problems of diagnosis and prevention that the disease posed. When it was time for Williams to return to the United States, she took a culture of the rabies vaccine with her. In her lab at the New York City Department of Health, she cared for the culture, coaxing it to grow. Eventually she had enough to produce vaccinations for fifteen people. Following Williams's interest in it, producing the vaccine became a major initiative in the United States.

Having handed off one part of the problem, Williams flipped back to studying detection. Rabies was maddeningly difficult to diagnose, and by the time scientists concretely pinpointed the disease in a patient, the opportunity for a vaccine had already expired. Rabies affects the nervous system and brain, so Williams started to look for flags the virus plants inside the body that might be used for early detection. And sure enough: Williams noticed that the virus was manhandling the structure of cells in the brain. It was major news, but Williams lost the headline again. While meticulously checking and double-checking her results, an Italian physician named Adelchi Negri independently discovered the cells. He beat her to the pages of a scholarly journal. Those rabies-affected cells are now called Negri bodies.

From rabies, Williams worked her way through research on venereal disease, eye infections, influenza, pneumonia, meningitis, and smallpox. Early on, her studies were powered by the drive "to find out about the what, why, when and where and how of the mysteries of life," she explained. "This trait had increased with the years, and finally had become a passion."

In 1934, Williams and nearly one hundred other workers were forced to retire by New York City mayor Fiorello La Guardia because they were over seventy years old. The mayor sent Williams packing, but word of her important contributions to bacteriology had been made clear to him. She was, he said, "a scientist of international repute."

ALICE BALL

1892–1916

CHEMISTRY • AMERICAN

JACK LONDON CALLED KALAUPAPA, A TILE OF LAND ON THE Hawaiian island of Molokai, "The pit of hell, the most cursed place on earth." On three sides, the area is surrounded by ocean. On the fourth, it's fenced in by a sheer two-thousand-foot cliff. It wasn't easy to get in. It was even harder to get out.

The area's occupants were struck with what was called the living death. Beginning in 1866 and lasting for eighty years, some eight thousand people with leprosy were ripped from their homes, arrested, and relocated to Kalaupapa, never to be seen again. For families, these departures were treated like deaths. A funeral was held, property was distributed, and families mourned the loss of a person still living. Sufferers were considered dangerous disease-spreaders and, without a cure, a lost cause.

Leprosy ravages the skin. It attacks the mucous membranes in the eyes, nose, and throat, and it goes after the peripheral nerves, located outside the brain and the spinal cord. The ability to feel pain goes, as do chunks of skin that develop into lesions. The damage is caused by a relative of tuberculosis. Although the disease is not as contagious as most think, to this day, doctors still don't understand how it moves among patients.

For hundreds of years, the closest thing to leprosy medication was an oil that came from the seeds of a chaulmoogra tree. People smeared it on the skin, swallowed it, and even injected it, but each delivery method was problematic. Rubbing it on like lotion didn't have any negative effects, but it also didn't do much good. The oil's acrid taste made swallowing it nauseating. When it was injected, the treatment just sat under the skin in a lump getting along exactly as you'd expect oil and water to. The injection became a subcutaneous snail; as it traveled, it burned. There were no good solutions.

But researchers were looking for one. A surgeon named Harry T. Hollmann, working out of Kalihi Hospital in Honolulu, just an island over from Kalaupapa, took particular interest in leprosy patients because when they fell ill, he was one of the doctors who treated them. Chaulmoogra oil was introduced in the Hawaiian Islands in 1879, and Hollmann was intrigued by its storied properties. Some patients really did seem to show improvement, but the benefits, as a whole, were scattershot. (One reason for the uneven theraputic effects was that not all oils peddled as chaulmoogra were the real deal.)

Hollmann was one of many scientists all over the world looking for a better way to whip the chaulmoogra oil into an injectable therapy for leprosy. The project needed a chemist, and that chemist was Alice Ball.

Ball was in her early twenties and an instructor at the College of Hawaii when Hollmann reached out. She had completed her undergraduate education at the University of Washington, earning a degree in chemistry in 1912 and then another in pharmacy in 1914. She lived in Hawaii as a child, when her parents moved from Washington State to Honolulu to take advantage of the warmer climate, hoping that it would ease arthritis's grip on Ball's grandfather. The relocation lasted a single year. When her grandfather died, the family returned to Seattle.

After her studies at the University of Washington, Ball published an article in the *Journal of the American Chemical Society* and then returned to Hawaii for a master's in chemistry. In 1915 she was both the first woman and the first African American to earn a graduate degree at the College of Hawaii. Ball continued on as an instructor at the school.

When she began her work with Hollmann, the chaulmoogra oil task was so knotty that it had already stumped many of his contemporaries. When a treatment is not water soluble, often scientists will coax it into its salt form, which can be absorbed by the body. But in the case of chaulmoogra oil, the salts would be so big that they'd act like soaps, which could badly damage a body's red blood cells. Ball had to find her own way to the solution.

Untreated, the oil was closer to honey than the stuff you cook with. Ball had to figure out a way to thin it out. She nudged it into forming a better relationship with water, so it would be absorbed rather than repelled. Ball treated the oil's fatty acids with an alcohol and a catalyst to kick-start the reaction to create a less viscous chemical compound.

With a bit more finessing, Ball became the first person in the world to successfully prepare a form of the oil that could be injected and absorbed by the body. In her formulation, there were no abscesses or bitter taste, and patients were able to get some relief. Ball, who fit the research in around her teaching schedule, made the breakthrough when she was just twenty-three years old.

Soon after solving one big problem, Ball was hit with another. At age twenty-four, in the middle of instructing a class, Ball may have mistakenly inhaled some chlorine gas. This time the reaction did not go in her favor. Chlorine interacts with water in the body, turning it to acid. Ball was flown back to Seattle in a last-ditch effort to save her, but ultimately the damage was too great and she died.

In 1918, two years after her death, an article in the *Journal of the American Medical Association* reported that seventy-eight patients with leprosy admitted to the Kalihi Hospital had been discharged—not back to Kalaupapa, but to their original homes. Ball's chaulmoogra oil preparation worked. For four years, not a single new patient was exiled to Kalaupapa, and other leper populations were let out on parole—thanks to a gender, race, and chaulmoogra oil barrier-breaking chemist.

GERTY RADNITZ CORI

1896–1957

BIOCHEMISTRY • CZECH

"AS A RESEARCH WORKER, THE UNFORGOTTEN MOMENTS OF my life are those rare ones, which come after years of plodding work, when the veil over nature's secret seems suddenly to lift and when what was dark and chaotic appears in a clear and beautiful light and pattern." These were Gerty Cori's words, first recorded for the radio series *This I Believe*, and played again at her memorial service in 1957.

Over the course of her career, Cori and her research partner/husband Carl lifted that veil repeatedly to reveal a dazzling string of discoveries, including processes as essential as how food fuels our muscles. When we talk about glycogen and lactic acid and how they relate to exercise, we're talking about a series of biochemical cycles that taken together are called the Cori cycle. The Coris were the first to bioengineer glycogen in a test tube, which was huge, considering that when they did it in 1939, no one had created a large biological molecule outside of living cells before. The Coris also ferreted out a whole slew of enzymes and then worked out how those enzymes control chemical reactions. Today, these discoveries are so fundamental to how we understand biochemistry that learning about them is standard in high school textbooks.

From 1931 until her death, Gerty Cori ran a lab at Washington University in St. Louis that was considered *the* epicenter of enzyme research. Scientists came from all over the world to work with her and her husband, and all in all the lab churned out eight Nobel Prize winners.

Of the couple, Gerty was the "lab genius" according to a colleague, keeping a sharp eye on the experimental data and commanding perfection. Her natural pace was breakneck, and she brought everyone else in the lab, including Carl, with her. She kept on top of current research, dispatching students to the library regularly to copy the most interesting articles. When she read something that jumped out at her, she'd race down the hallway to Carl's office to discuss it with him. She smoked furiously, and evidence of her frenetic activity could be seen in a flying trail of ashes.

Though she could be tough on colleagues, Gerty's command of the lab had a lot to do with keeping experimental conditions ideal—and just how flat-out thrilled she was to be doing important work in biochemistry. If she took someone to task for a mistake, her response came from disappointment; she would have to delay the fun stuff for a day.

At the core of so many of the lab's discoveries was Gerty's partnership with Carl, which was forged during medical school at the University of Prague in anatomy class, of all places. Even before they were married, they published together. Once they wed in 1920, they wanted to continue. However, there were powerful political, social, and geographic forces in Eastern Europe working against them. Gerty converted to Catholicism to marry Carl, but even so, anti-Semitism was so acute that his family worried about his job prospects suffering because of her Jewish ancestry. Furthermore, the borders of what was once Austria-Hungary were in flux. At one point, Carl and some friends dressed up in workmen's clothes to break down a lab in Czechoslovakia in secret, only to reassemble it again in Hungary, the home of the lab's founder.

The Coris decided that the best opportunities for them were likely abroad. Carl was offered a position in Buffalo, New York, at a research center tackling malignant diseases. Gerty stayed at a children's hospital in Vienna, where she dug into congenital thyroid deficiency, until she got the thumbs-up from Carl that he had lined up a position for her at the research center in Buffalo as an assistant pathologist. Six months after he arrived, she followed him to the United States.

In Buffalo, they fought with administration for their autonomy. When a director added his name to their papers without reading the contents, the Coris removed it before sending their manuscripts for publication. When the director pushed his theory that cancer was the work of parasites, Gerty refused to play along, and was nearly fired for it. The director demanded that the only way she would be able to hold on to her position at the center was to stick to her own lab space and quit collaborating with Carl. Naturally, they snuck around, peeking at each other's slides and discussing results.

Soon enough, they were back in the swing of things, traveling deep into the study of glycogen together. In nine years, they published fifty papers and mapped out the general structure of the Cori cycle.

When it was time for them to move on, Cornell University, the University of Toronto, and the University of Rochester Medical School all called Carl. Gerty, however, was the deal breaker. She was scolded by a school courting Carl, told that by requiring a position at the universities she was torpedoing his career. What the schools didn't understand was that the Coris did their best work together. There was sexism at play, certainly, but nepotism rules also made hiring a spouse difficult. With so many Americans out of work during the Great Depression, two family members working at the same university was seen as an unfair advantage.

Washington University School of Medicine in St. Louis found a Gerty-sized loophole, in that the school was part of a private institution, not a public one. Carl was brought on as a research professor and Gerty as a research associate. Though their titles were tiered, the Coris always acted like they were equals.

During work hours the two talked to each other constantly about their research, but they were close outside the lab, too. At home, they skated and swam and hosted parties and tried to avoid discussing ongoing experiments in their limited off-hours. When they were at the university, an hour-long lunch break became a Cori-led story time, during which they regaled their colleagues with discussions of far-reaching passions. The talks covered everything from wine to research reports to whatever they'd been reading for pleasure.

When the hour was up, the team was back at top speed. In 1936, the Coris figured out how the body breaks down glycogen into sugar. The final years of the 1930s were dedicated to tracking down new enzymes and sussing out their purpose. When the Coris netted phosphorylase, it was the first time scientists had zoomed in to observe the molecular workings of carbohydrate metabolism.

Test tube glycogen gave the Coris—and the larger research community—another high. Carl made quite the spectacle by whipping the molecule up in front of a conference audience and then passing the test tube around for all to ogle.

Pressure from outside finally bumped up Gerty's official status at the university. An offer from Harvard and the Rockefeller Institute to make them both professors was turned down only when Washington University countered by promoting Gerty.

According to the science journalist Sharon Bertsch McGrayne, "The lab made so many discoveries so quickly during the late 1940s and early 1950s that Carl worried a bit." It wasn't luck that brought them so many successes; it was hard work. Gerty was a fixture in the lab every single day.

Gerty found out that she and Carl had won a Nobel Prize

"for their discovery of the course of the catalytic conversion of glycogen" in 1947, the same year she found out she had a rare form of anemia, which would kill her a decade later. Wherever promising treatments were offered, the Coris went. They traveled all over the world in an effort to find something that would improve her health.

Gerty kept her illness to herself, but others could see traces of it in her routines. The Coris moved a cot into her lab so she could rest; blood transfusions zapped her energy, and she became more frequently frustrated. In one illness-related incident, Gerty fired the nurses Carl had hired to help her. Though she could no longer be the Gerty who would jet down the hall or leap in excitement at a breakthrough, she kept working with the same fervor that had always defined her. When she could no longer make the trek from one room in the lab to the other, Carl scooped her up and carried her, working together until the end.

HELEN TAUSSIG

1898–1986

MEDICINE • AMERICAN

HELEN TAUSSIG STUDIED THE HEART, BUT SHE COULD NOT hear it. That *thu-thud, thu-thud* sound started fading when she was around thirty, the most fundamental indicator of life eluding failing ears. As a stopgap, she modified her stethoscope with an amplifier. But as her hearing deteriorated further, Taussig began feeling for the heart's beat instead of listening for it. She read the rhythm like Morse code, interpreting the signal for an organ's abnormalities. Supplemented with blood pressure results and electrocardiogram images, Taussig pieced together a series of clues in order to come to a plausible diagnosis. She called this triangulation process "the crossword puzzle."

Taussig was a founder of pediatric cardiology. When she entered the field around 1930, it was considered a dead-end specialty. Before open-heart surgery, physicians could diagnose heart abnormalities, but they couldn't do anything substantial to treat them. Pediatric patients routinely died. When they did, the autopsy would provide the crossword puzzle's final clue.

Without a clear understanding of where it would lead, Taussig gathered data about her patient's health and heart in order to formulate a diagnosis. Gathering data wasn't a solution to the heart abnormalities she saw in patients, but after more than a decade of observation and testing, Taussig amassed the most comprehensive catalog of congenital heart defects and their indicators ever compiled.

Taussig had plenty of experience in perseverance. Her mother died when she was eleven, and as a child, she worked extra hard in her classes to compensate for her dyslexia. Taussig found her way to cardiology despite being edged out of programs at three—*three*—universities: Harvard, Boston University, and Johns Hopkins. At Harvard, the rejection was particularly harsh. Because of her gender, Taussig wasn't allowed in Harvard's medical school, so she inquired about its newly opened School of Public Health, which overlapped with medicine and admitted women. The answer surprised her: women could attend, sure, but their efforts wouldn't earn them a degree. "Who is going to be such a fool as to spend four years studying and not get a degree?" Taussig asked. The dean replied, coolly, "No one, I hope." She did not enter the program.

It was in this way that Taussig ping-ponged from medical school to public health to cardiac research to pediatrics to pediatric cardiology. Looking back, she saw the places that rejected her as a route to greater opportunity. In 1930, she finally found her place as the director of pediatric cardiology at Johns Hopkins's Children's Heart Clinic. At first it was a lean operation. Taussig got assistance from a social worker and a technician, who jumped in to answer phones and file paperwork when needed. In the early days, there was a perception that by simply treating patients, Taussig was taking them from medical students, who might learn something from the meeting. So Taussig saw what patients she could and gathered data without a clear idea of how she would eventually use it.

A groundbreaking surgery opened up bright new possibilities for her patients. In 1939, a Harvard surgeon performed a procedure aimed at fixing ductus arteriosus closure. The ductus arteriosus is an opening in the heart that joins two important blood vessels. In the womb, it's supposed to be open, but as a baby's lungs begin to fill with air, the tunnel joining the two arteries should close. If the hole doesn't clamp shut automatically, babies get too much blood to their lungs, which can lead to congestive heart failure and oxygen-poor blood flowing through the body, turning patients blue. The surgeon developed a procedure to close the hole manually.

By this point, Taussig had been hard at work for almost a decade. She saw ductus arteriosus patients, but she also saw patients with multiple heart abnormalities who seemed to benefit from an open ductus. In some cases, the opening indirectly shuttled enough blood to the lungs to keep the patient alive. Taussig thought, perhaps surgery could open the ductus as well?

Taussig floated her idea to the Harvard surgeon. "Madam," he replied, "I close ductuses. I don't create them." Other surgeons were similarly skeptical, and colleagues expressed annoyance that she wouldn't drop the idea. Finally, after two years of campaigning, Taussig convinced Johns Hopkins's brand-new chief of surgery, Alfred Blalock, to get on board. Blalock asked Vivien Thomas, a technician at Hopkins's research lab, to figure out what surgical steps would be needed to nail the procedure. In 1944, with instruction from Thomas, Blalock performed the first successful Blalock-Taussig shunt on a fifteen-month-old girl. For the third patient, the surgery caused an immediate change in the child's appearance. "I suppose nothing would ever give me as much delight as seeing the first patient change from blue to pink in the operating room . . . bright pink cheeks and bright lips," Taussig remembered fondly. "Oh, what a lovely color."

So began a whole new era of pediatric cardiology. All of a sudden, there was a rush to Taussig's quiet specialty. She remembered it like this: "Dr. Gross [of Harvard] unlocked the gate . . . I opened it; Dr. Blalock and I galloped in, quickly followed by a stream of patients, surgeons, cardiologists, and pediatricians." In 1947, as a culmination of twenty years of research, Taussig literally wrote the textbook on congenital heart defects.

Taussig felt strongly that the new droves of pediatric cardiologists shouldn't have to cobble together an education as haphazardly as she'd had to. With funding from the National Institutes of Health and the Children's Bureau, pediatric cardiology got an official training program at Johns Hopkins with Taussig at the helm. Over the years, Taussig taught some 130 young doctors how to succeed in the field she'd started. She emphasized that patient care was an essential component of clinical training. She insisted physicians treat children as children, and not like sick kids. Compassion and patience were essential in dealing with both the patients and their stressed-out families. Her students, so taken with her verve for the human heart, proudly called themselves "the knights of Taussig."

By the time she died in 1986, she had published forty papers post-retirement, served as the American Heart Association's first female president, and received the Presidential Medal of Freedom, given to her by President Lyndon Johnson. Taussig was also instrumental in convincing the Food and Drug Administration to block a medicine that she believed (rightly) caused birth defects.

Over the course of her career, Taussig did an extraordinary amount for the heart. But she never forgot what it felt like to have her hands on it. "You have your sadnesses as well as your successes," Taussig admitted. "One reads all about the successful operation, but not about the unsuccessful ones, the sorrow and background of hard work. On the whole, though, I think I've done more good than harm."

ELSIE WIDDOWSON

1906–2000

NUTRITION • BRITISH

WITH A PILLOW ON HER LAP AND A SYRINGE IN HER LEFT HAND, Elsie Widdowson injected a cocktail of iron, calcium, and magnesium into her right arm. Going into the experiment, Widdowson and her research partner, Robert McCance, assumed that iron was excreted; however, based on their experiments in 1934, they realized that iron was instead absorbed.

When Widdowson began studying it in 1933, nutrition was still an emerging field called "dietetics." She entered into the discipline on the recommendation of an advisor, several years after earning her PhD. Widdowson had taken a position in the plant physiology department at Imperial College, London, where she monitored changes in apple carbohydrates. The job required twice-monthly apple-picking field trips to Kent, a county southeast of London, so Widdowson could monitor the fruit's composition at different stages in its life cycle, from its first blossom to storage.

After the study was over, she dabbled in biochemistry at Middlesex Hospital's Courtauld Institute. Widdowson didn't dislike apples, but she hoped to get closer to research that would more directly benefit humans. She found a position at King's College Hospital in London in 1933.

Before officially meeting McCance, she was aware of him. He was the scientist cooking up slabs of meat in order to learn about their chemical composition. The scientists had fruit research in common. When the pair finally got to talking, Widdowson informed McCance that he'd made a significant error in one of his assessments. Because he had failed to take into account a change in fructose, his carbohydrate numbers were too low; the study was wrong. McCance suggested they join forces. They would remain research partners until McCance's death in 1993.

McCance was already well on his way to having a few major food groups analyzed; he'd nearly knocked out the nutrition information for meat, fish, fruits, and vegetables. On an outing with her family in 1934, an idea suddenly struck Widdowson. Those initial categories were important, but why not go all in? They should analyze everything—sweets, dairy, cereals, beverages—*everything.*

The Chemical Composition of Foods was published in 1940. With fifteen thousand values, it was the first comprehensive compendium of nutrition information for cooked and raw foods . . . well, ever.

Meanwhile, Widdowson and McCance maintained a steady stream of side projects. In nutrition research in the mid-1930s, there was still just so much unknown. In one study, Widdowson and McCance wanted to learn about how salt deficiency affects the body. They rounded up some healthy (but reluctant) subjects and put them on a salt-free diet for two weeks. Each participant agreed to spend two hours per day in a podlike warming contraption that forced perspiration. With a long lab coat and her trademark Princess Leia braids, Widdowson hosed down subjects and their plastic pod sheet after the sessions. The researchers analyzed the runoff for salt content. When sufficiently salt-depleted, the weakened participants were subjected to an array of tests, particularly ones testing for kidney function.

Widdowson and McCance were the first to show just how important fluid and salt are for body function. Hospitals now keep a close watch over these levels, especially in the case of kidney disease, heart attack, or diabetes.

Much of Widdowson's work during the 1940s responded to

the urgent nutritional needs of populations as a result of World War II. It was during this time that Widdowson and McCance earned their titles as creators of "the modern loaf," a loaf of bread enriched with calcium. With meat, sugar, and dairy in limited supply, the British government was concerned about having enough nutritious food for its citizens. Widdowson and McCance predicted that their compatriots would do just fine eating a diet composed of a few staples in abundant supply: cabbage, potatoes, and bread. To test the veracity of their hypothesis, Widdowson and her colleagues all gave the simple, color-free diet a try for three months. The study ended with a rigorous two-week hike-a-thon in England's Lake District. McCance rode his bike there—a two-plus-day trip—and Widdowson drove up with other colleagues. They hiked every day and took notes on how they fared. On one of the days, McCance traversed thirty-six miles, powered mainly by cabbage, potatoes, and bread. The simple diet was deemed a grand success, save for a bit of missing calcium, which was added by supplementing flour with chalk. When rationing went into effect during World War II, the government gave their pared-down diet a big public relations push. Although food was scarce, at no other time in English history had the population eaten healthier.

After the war ended, Widdowson traveled to Germany to help find solutions for the malnourished. Bread, again, was of particular interest. Widdowson went from orphanage to orphanage in the thick of winter to recruit locations for a bread study that would compare different refinement processes. For eighteen months, Widdowson monitored orphan children's height and weight and matched the results against the type of bread they consumed. The level of enrichment in the bread flour didn't seem to make a difference, but while she was there, Widdowson noticed sizable changes in the children's weight and growth that seemingly had nothing to do with the loaf. At one orphanage growth slowed dramatically at about the same time as the children's weights at the other test site suddenly took off. Concerned as to the cause, Widdowson launched an examination into what external factors might be at play. The process of elimination pointed to a particularly cruel housemother who had transferred from the first test location to the second. Where she was present, height and weight gains stagnated. Widdowson concluded, "Tender loving care of children and careful handling of animals may make all the difference to the successful outcome of a carefully planned experiment."

Widdowson's hands-on approach to research went awry only a few times. When she and McCance carried out another round of self-injections, they wound up on the floor, writhing with fever, tremors, and body aches. A colleague had to take them home to nurse them back to health. Nevertheless, Widdowson and McCance carried on, collecting their samples even through cold sweats. "A slight accident," they conceded.

Widdowson loved really digging her nails into a problem, whether that meant throwing a dead baby seal in the trunk and driving it from Scotland to Cambridge to analyze its fat content or running back and forth through an airport's metal detector to figure out what on earth could be causing all the beeping. The allure of the experiment was just too enticing to resist. Her curiosity did the body good.

VIRGINIA APGAR

1909–1974

MEDICINE • AMERICAN

WHETHER SHE WAS BICYCLING WITH A COLLEAGUE'S CHILD, cheering at a baseball game, or taking flying lessons, Virginia Apgar always kept the following things on her person: a penknife, an endotrachial tube, and a laryngoscope, just in case someone needed an emergency tracheotomy. Even when she was off-duty, she was on: "Nobody, but nobody, is going to stop breathing on me."

Apgar was one of the earliest medical doctors to take up anesthesiology. She was fast talking and fast thinking, an endless spring of energy. Growing up in New Jersey with an amateur inventor and scientist as a father and a chronically ill brother, Apgar quipped that her family "never sat down." She didn't, either. In college, while earning top marks studying for a zoology degree, Apgar churned out articles for her college newspaper, became a member of seven sports teams, acted in theater productions, and played violin in the orchestra. After the stock market crash of 1929 hit her family hard, Apgar picked up a collection of odd jobs, including one catching stray cats for the zoology lab. "Frankly, how does she do it?" asked the editor of her high school yearbook. The question could be applied to any stretch of her life.

There were some things she didn't make time for, namely, bureaucracy and red tape, which she'd walk over if she felt like they were stopping her from helping a patient or doing the right thing. If an elevator frightened a child, she'd scoop him or her up in her arms and take the stairs. During her medical school residency, Apgar worried that she might have made a mistake that contributed to a patient's decline. She asked for an autopsy, but it wasn't granted. Uncovering the truth became an irrepressible need, so she snuck in and reopened the incision herself. Apgar immediately reported her error to her superior.

She had no tolerance for insincerity or deception. Her own example of openness—an ability both to admit her failures and to adapt as anesthesiology changed—was instrumental in helping the discipline move forward. In fact, it was her flexibility that got her to anesthesiology in the first place.

When Apgar began her internship in surgery at Columbia University in 1933, she was one of only a few women studying it in the country. She worked under the chair of surgery, who recommended she shift focus and join the emerging field of anesthesiology, which, at that point, wasn't even considered a medical specialty. For her advisor, the recommendation was self-serving. He admired her abilities and he spotted a need. At the time, if a patient required anesthesia, a nurse stepped in. But as surgeries became more complicated, Apgar's advisor realized that anesthesia would have to keep pace with highly skilled practitioners talented and driven enough to forge the way in a rapidly emerging field.

Apgar spent a year away from Columbia to train. When she returned in 1937, she laid out a plan for how the Division of Anesthesia would function within the Department of Surgery at Presbyterian Hospital. She asked for a title (director), suggested an organizational structure, and mapped out how to establish residencies and bring in more specialists without displacing the nurses already working in anesthesiology. For eleven years, Apgar headed the division, training medical students, recruiting residents, and doing research. She played a major part in helping the specialty grow, but when the division became a department, a male colleague was given the position of chair.

Apgar turned her focus to babies. While administering gas to women in labor, she noticed a curious lack of data. The statistics she did have were puzzling. Thanks to hospital deliveries, more mothers and babies were living through birth, but for newborns, the first twenty-four hours remained particularly perilous.

When Apgar looked into the issue, she saw something striking: infants weren't being examined right after birth. Without an immediate appraisal, doctors were missing signs that a baby was, say, starved for oxygen, a factor in half of newborn deaths. Furthermore, Apgar realized that a set of standards to compare newborns against simply didn't exist. If the mother was given drugs during labor, sometimes her baby would take one breath and not take another for several minutes. Did that count as breathing or not breathing? It depended on the delivering physician. Apgar stated what now sounds like the obvious: a baby shows clear signs when it's struggling, and all babies should be monitored for these red flags.

How would one go about doing a quick, standardized assessment of a newborn, a medical resident asked Apgar. "That's easy," she replied, grabbing a nearby piece of paper. "You would do it like this."

The scoring system would cover five major areas requiring a physician's attention: heart rate, respiration, reflex irritability, muscle tone, and color. Each was rated on a scale of 0 to 2. Almost immediately, Apgar and some colleagues deployed the system to find connections between scores and a baby's health. Low scores, they found, send up a signal that there are carbon dioxide and blood pH problems. In the case of an overall score of 3 or less, the baby was almost always in need of resuscitation.

A single baby's score was powerful, but the effect of analyzing thousands of them was like a field of fallen leaves suddenly organized by color, all these little bits of evidence sorted to reveal their common cause. Lower scores correlated with certain methods of delivery and types of anesthesia given to the mother. Before this small and efficient scoring system, doctors just didn't see these connections—or didn't have consistent enough data to prove them. The scoring system became a foundation for better public health statistical models. It began to spread from New York to hospitals across the country.

When it reached Denver, the score finally got its famous name. In 1961, nine years after its initial presentation, a medical resident came up with a catchy mnemonic device:

A- Appearance (Color)
P- Pulse (Heart rate)
G- Grimace (Reflex irritability)
A- Activity (Muscle tone)
R- Respiration

And there it was, the Apgar score. Apgar (the anesthesiologist) loved it.

Meanwhile, data had been pouring in and Apgar didn't feel adequately equipped to deal with it. Always open to the things that would make her a better doctor, Apgar took a break from her duties at the hospital to pursue a master's degree in public health. Spotting an opportunity, the National Foundation–March of Dimes swooped in and made her an offer.

Here, as ever, her curiosity drove her decision making. Enticed by the idea of a midlife career change, Apgar finished up her degree and leapt into her new role as chief of the organization's new Division of Congenital Malformations.

For fourteen years, Apgar flew across the country, spreading information on the reproductive process and trying to dispel the stigma surrounding congenital birth defects. Her quick and witty personality made her a favorite of TV hosts and the patients she visited. As it was often said, she was a "people doctor," as quick to connect with patients as with viewers and absolutely everyone else she met. "Her warmth and interest give you the feeling that her arms are around you, even though she never touches you," said one volunteer who worked with her. The foundation doubled its income while she was in a leading role.

Apgar worked with people, flew planes, and cheered for baseball games (her colleagues and friends panting behind her) until her health stopped her. Although she died in 1974, her score is still around. It has protected babies all over the world for the better part of the last century.

DOROTHY CROWFOOT HODGKIN

1910–1994

BIOCHEMISTRY • BRITISH

IN THE CAVELIKE ACCOMMODATIONS IN THE BASEMENT OF the Oxford University Museum, electrical cables hung from the ceiling like a high-voltage canopy of Christmas lights. A single Gothic window graced the lab space, but it was mounted so high that taking advantage of its light required a staircase. In the twenty-four years that Dorothy Hodgkin ran her X-ray crystallography lab from the museum, at least one person was zapped with 60,000 volts—and thankfully, not fatally. The lab was underfunded and Hodgkin underappreciated, but she made do. Even in the paltry conditions, Hodgkin's masterful abilities launched her to the top of her field.

X-ray crystallography became a discipline in 1912 when Max von Laue discovered that X-ray diffraction patterns can tell scientists quite a bit about a molecule's atomic structure. The process starts with molecules all organized in a uniform, recurring pattern, called a crystal. When X-rays are pointed at crystals, the molecules cause the X-ray to diffract, and the resulting design is captured on photo plates. The pictures are chock-full of clues that can lead researchers to the 3-D structure of the molecule. To decode them before computers was an especially gnarly task—one that could take years of computational muscle and exceptional patience. Hodgkin was a pro.

In the early 1930s, during the first years of Hodgkin's career, cracking even the simplest crystal's code took tens of thousands of mathematical calculations carried out on a hand-adding machine. The equations were used to build what's called an electron density map, which looks like a topographical map but shows instead where the crystal's electrons are most concentrated. The whole process, from X-ray to structure, could take months or even years.

In 1936, plowing through calculations got a bit easier when Hodgkin became the proud owner of two boxes packed with 8,400 thin pieces of paper. These so-called Beevers-Lipson strips were like a card catalog for crystallographers. From top to bottom, they were filled with meticulously ordered trigonometric values, which cut down the time Hodgkin spent working out the math.

When she began decoding the cholesterol molecule in the late 1930s, most of her peers said it couldn't be done with crystallography. But Hodgkin, whom a friend affectionately called the "gentle genius," beamed X-rays at the cholesterol crystal and started punching that adding machine. Where traditional chemists had failed, the crystallographer succeeded.

Word of her incredible electron density map decoding skills spread, and Hodgkin found herself a magnet for unsolved crystal structures. When someone needed a molecule's structure worked out, they dropped off the crystal sample at Hodgkin's. Over the years she was sent some doozies, penicillin among them.

Penicillin had already shown its ability to prevent bacterial infection in humans by 1941—an extraordinary boon during wartime. By understanding its structure, scientists hoped to help drug developers mass-produce it. However, the molecule evaded scientific understanding in a variety of ways. First, American and British crystallographers were unknowingly working on penicillin crystals of different shapes. Nobody knew penicillin crystals could have those variations. Furthermore, because of the way the molecules were layered, the photo plates didn't present a very clear picture, either.

Finally, as if there weren't already enough challenges, Hodgkin and her Oxford graduate student were endeavoring to work out the structure of the penicillin molecule without any knowledge of its chemical groups. Hodgkin joked that it looked like "just the right size for a beginner."

Hodgkin's decoding work revealed that penicillin's parts were bound together in an extraordinarily unusual way. One chemist was so taken aback that he wagered his whole career against her findings, swearing he'd become a mushroom farmer should her assertions about its structure be true. (Despite her verified results, the naysaying scientist didn't become a mushroom farmer.) When Hodgkin realized she had the final answer to the penicillin problem in 1946, she flitted around the room in childlike celebration. It had taken her four years. The discovery ushered in new, semisynthetic penicillins and their widespread deployment.

Her success notwithstanding, it would be eleven more years until Oxford would make her a full university professor. For an upgraded lab space, she'd have to wait twelve.

Her next massive molecular puzzle had six times more nonhydrogen atoms than penicillin's total number. Overturning expectations was a Hodgkin hallmark. Although other scientists declared that B_{12} was unsolvable by X-ray crystallography, Hodgkin gave it a shot.

For six years, Hodgkin and her team took some 2,500 X-ray photos of B_{12} crystals. Processing the images was beyond anything the Beevers-Lipson strips could handle. Luckily, Hodgkin had a computer programmer on her side. The University of California, Los Angeles, had just gotten a new computer specifically programmed to tackle crystallographic calculations, and the student programmer, a chemist, just happened to be visiting Hodgkin's lab at Oxford for the summer. When the student returned to UCLA, Hodgkin mailed him bundles of information on B_{12}, and he would send back the computer-processed results. The work was long, difficult, and very challenging. When there were mistakes, like one that rendered an atom ten times larger than its actual size, Hodgkin told her Southern California programmer to cheer up. Through the entire process, he never once saw her lose her cool.

Eight years after she'd commenced work on B_{12}, Hodgkin successfully nailed down its 3-D blueprint. According to a British chemist, if her work on penicillin "broke the sound barrier," B_{12} was "nothing short of magnificent—absolutely thrilling!"

"For her determinations by X-ray techniques of the structures of important biochemical substances," Hodgkin was awarded the Nobel Prize in Chemistry in 1964.

Hodgkin was always kind and gracious but also firm when she needed to be. Underestimating her never went well. Even in her old age, she continued to surprise people. With crippling rheumatoid arthritis and a broken pelvis, she continued to jet to Moscow and elsewhere to attend conferences on science and peace.

GERTRUDE BELLE ELION

1918–1999

BIOCHEMISTRY • AMERICAN

GERTRUDE ELION NEVER FORGOT THE ONES SHE LOST: HER grandfather from stomach cancer when she was a teenager, her fiancé from a sudden infection in his heart, a patient from leukemia, her mother from cervical cancer. She held on to the very real pain of their passing, the losses serving as a constant reminder that every atom substituted and drug synthesized might make a difference. Her grandfather's death "was the turning point," she admitted. "It was as though the signal was there: 'This is the disease you're going to have to work against.' I never really stopped to think about anything else. It was really that sudden."

Though her purpose was singular, her path to pharmaceutical research was not. The first hang-up was funding a higher degree in chemistry. She applied to fifteen schools, and not a single graduate program would offer her any kind of financial assistance. During the Great Depression, what money the schools had went to men. It was the same in the job market. One employer worried that with no other women in the lab, Elion would be a "distracting influence."

To get herself closer to the chemistry she loved, Elion adapted, pulling together a hodgepodge of employment. She signed up for secretarial school and taught nursing students biochemistry. When she ran into a chemist at a party, she offered to work for free. Eventually she scraped together enough money to fund one year of graduate school at New York University. Elion supported herself by picking up a job as a doctor's office receptionist.

Elion's first full-time position in a lab was quality control for a line of grocery stores. She tested the acidity in pickles and made sure spices were fresh. She took what she needed from the position and then called it quits, telling her manager matter-of-factly, "I've learned whatever you have to teach me, and there's nothing more for me to do. I have to move on."

Her father noticed "Burroughs Wellcome Company" on a pill bottle and suggested that she apply for a job, since the company was located nearby, in Westchester County, New York, just eight miles from her home. Burroughs Wellcome gave scientists the space, freedom, and financing to chase down drug-related solutions to any serious medical problem they desired. When Elion showed up for an interview, it was luck that she landed in the company of George Hitchings, who was working on just the types of problems that Elion wanted to tackle.

In 1944, Elion was hired by Hitchings, who was interested not only in drug development but also in how the medical research community was conducting it. Trial-and-error drug development was the norm, but Hitchings believed the method was a little like grabbing at a solution hidden somewhere inside a paper bag. Why couldn't they learn about new drugs with a methodical, scientific approach that incorporated knowledge of applicable subjects like cell growth?

Hitchings sent Elion off to explore adenine and guanine, the so-called purine bases in nucleic acid. (They're the A and the G in ACGT—the building blocks that make up DNA.) Cells need nucleic acids to reproduce, and tumors, bacteria, and protozoa need a lot of them to spread. So Hitchings figured that really getting to know these little-understood acids might allow the research team to develop a biochemical wrench to throw at diseases to stop them from spreading.

Thrilled to finally be doing work that satisfied her, Elion stayed late, went into the lab on weekends, and carried out her experiments merrily—even when the floors were a sizzling 140 degrees, a by-product of the baby food dehydration plant downstairs, which ran even during New York's muggy summer. She enjoyed the job so much that when she opted to spend one whole weekend at home and put work on the backburner, her mother worried that there must be something wrong with her.

Elion was in her element. She ripped through studies of organic chemistry, biochemistry, pharmacology, immunology, and virology. However, she still yearned for a PhD. For a time, Elion took PhD classes in her off-hours, but she ended up getting forced out of the program. Work on your PhD full-time or get out, the dean demanded. She chose her job over higher education. "Oh, no, I'm not quitting that job," she explained to the dean. "I know when I've got what I want." (Elion never did complete her PhD, though George Washington University gave her an honorary one.)

Elion stayed put because she was just having too much fun and too much success waging a battle against the illnesses that kill us. Take her accomplishments in 1950. Elion synthesized two effective cancer treatments. Remember those purine bases? Well, she developed a compound called diaminopurine that used them to disrupt leukemia cell formation. On animals, diaminopurine worked such wonders that Sloan-Kettering Memorial Hospital in New York City tried the medication on two severely ill leukemia patients. One patient's recovery so transformed her that, for a while, doctors thought she might not even have had leukemia in the first place. The patient stopped taking the medication, got married, and had a child. Elion also developed another compound that improved the life expectancy of leukemia patients.

Both medications were major breakthroughs in cancer treatments, but their effectiveness only took patients so far. When the leukemia-patient-turned-mom relapsed and died two years after receiving the medication, Elion was completely torn apart. Even after decades had passed, the case would still move her to tears.

As the years went by, it was almost as if Elion would skip through areas of disease research and then invite the world to come along. In the early 1950s, she started studying drug metabolism before she'd heard of anyone else doing it. That pair of cancer drugs? Her work kicked off a whole new wave of leukemia research. Then, in 1978, Elion and her research partners completely flipped the way scientists thought about viruses.

Antivirals were not thought to be very accurate attackers. Scientists believed antivirals would aim for a virus's DNA but would smash into the DNA of healthy cells, too. Elion's antiviral research, however, got a promising start. When Elion sent a sample of some early work to a lab to be tested, the response she got back was encouraging. "This is the best thing we've seen. It's active against both the herpes simplex virus and the herpes zoster virus." For four years Elion and her team fine-tuned the compound to bump up its effectiveness and nail down its metabolism. The secret? What Elion and her colleagues developed was a near twin, something so similar that the virus itself activated the assassin that would take it out. The drug was unveiled at a scientific conference in 1978 and was immediately heralded as a major breakthrough, one that would change the way scientists approached antivirals.

Elion lived for these highs. But better than solving any chemical puzzle was how her work touched people. In 1963, she watched a medication she'd helped develop clear up a watchman's painful gout, and in 1967, the first heart transplant took place thanks to an immunosuppressant she worked on.

"For their discoveries of important principles for drug treatment," Elion and Hitchings won the Nobel Prize in Medicine in 1988. After she received the award, letters poured in from people expressing their gratitude. Someone recounted a story of a son's terminal reticulum cell sarcoma reversed; another wrote about a daughter with herpes encephalitis whose life was saved. Someone else's eyesight was protected from a case of severe shingles, thanks to the medicine Elion had helped pioneer. She may not have been able to do anything about the deaths of her loved ones, but, explained a research vice president at her old institution, "In fifty years, Trudy Elion will have done more cumulatively for the human condition than Mother Teresa."

JANE WRIGHT

1919–2013

MEDICINE • AMERICAN

JANE WRIGHT BECAME THE DIRECTOR OF THE CANCER RE-search Foundation at Harlem Hospital in 1952 before she reached her mid-thirties. "Lollygagging," says her daughter, was just not in her nature. Weekday or weekend, at home or on vacation, on her way to the lab, in a restaurant, or boating in Michigan, Wright was up early and dressed to the nines. She shot up through the ranks of Harlem Hospital and then to the top of New York Medical College. By 1967, there was not another African American woman in a nationally recognized medical institution with a more prestigious position. For her pioneering work in cancer treatment, Wright was known as "the mother of chemotherapy."

But she was nearly known by another name. When Jane Wright first started at Smith College in Northampton, Massachusetts, she wanted to be a "renowned artist." But some advice from her father convinced her to switch from painting to premed. Wright came from a long line of high-profile doctors. Her grandfathers, Ceah Ketcham Wright and William Fletcher Penn, were both doctors. The latter was the first African American to graduate from Yale's medical school. Wright's father, Louis Tompkins Wright, was a highly respected surgeon and cancer researcher. Perhaps, her father suggested, she should major in something with a more secure route to employment. From Jane Wright's perspective, the challenge had been thrown down.

Wright entered premed extremely determined, not only to do well in her classes but also to keep up everything else she enjoyed. Wright balanced her medical school responsibilities with swim team practices and editing the yearbook. She graduated from Smith in 1942 and went on to receive an MD from New York Medical College in 1945. She was a bottomless well of energy. Her supervisor at Bellevue Hospital called her the most promising intern to have worked with him. She may have been a talented artist, but Wright was quickly proving herself an excellent doctor, too.

As she trained, her father's stellar reputation in the medical community set a constant example for her own studies. That is to say, like a billboard perpetually fifty feet ahead, the accomplishments of Louis Tompkins Wright could be both motivating and tiresome. To give you a sense of his status—and of hers—Wright was interviewed by the press following her graduation from medical school. "His being so good really makes it very difficult," she admitted. "You feel you have to do better. Everyone knows who Papa is."

Seeing her promise, in 1949 Louis Tompkins Wright invited his daughter to come work with him at Harlem Hospital's Cancer Research Foundation, an organization he'd recently founded. Together they would dive into what Wright would eventually call "the Cinderella" of cancer research: chemotherapy.

When Wright and her father started working together, physicians and scientists were just—*just*—starting to make some headway in finding treatments that could affect spreading cancer cells. In 1945, the director of cancer research at Columbia University described the scope of the task: "It is almost, not quite, but almost as hard as finding some agent that will dissolve away the left ear, say, yet leave the right ear unharmed—so slight is the difference between the cancer cell and its normal ancestor."

Scientists had made some headway with a chemical related

to mustard gas, called nitrogen mustard. Using a chemical weapon to treat cancer wasn't an obvious choice. But an unrelated tragedy—a navy ship that sank in 1943 while carrying mustard gas—gave scientists a tip that something in the chemical might work for cancer patients. When the ship went down, the mustard gas leaked. Many of the soldiers exposed died. During the gassed soldiers' autopsies, it was discovered that the chemical annihilated white blood cells—the ones that protect the body against infection. Those cells are also the ones that grow cancer in leukemia patients. In 1946, the first cancer patient injected with nitrogen mustard saw an improvement.

So for three years, until his death in 1952, Wright and her father tested drugs that might force leukemia into remission, trying to differentiate the left ear from the right. When her father passed away, Wright stepped up as the head of the research group her father had founded. She was thirty-three.

Over the course of her career, Wright steadily advanced the effectiveness of cancer therapies. One of her most important insights was that one magical solution wasn't going to swoop in and cure cancer for everybody. Say researchers find a really good drug cocktail to beat back breast cancer. When that therapy is applied to a different type of cancer—perhaps lung or colon—it might flop. Not even two cases of the same type of cancer can reliably be treated identically. If cancer cells spread quickly and the treatment is off, patients lose crucial time to dead-end therapies.

For twenty-two years starting in 1953, Wright worked on made-to-order solutions. When someone came in sick, Wright took a sample from the patient's tumor so she could grow those cancer cells in a lab. Wright used the sample—not the patient—to test-drive a drug's ability to vanquish the disease. An effective drug mixture in the lab meant it was worth advancing the therapy to the body. The approach didn't waste a patient's time on ineffective drugs, and it was faster and more personalized than using mice as proxies.

Wright also broke new ground on drug delivery. When cancer showed up in hard-to-reach places like the kidneys, surgery was often the default method of tumor extraction. Wright developed a system that would funnel drugs to a targeted area via catheter.

Extremely determined but always modest, many of Wright's accomplishments were unknown to her own daughters until her death, when friends and colleagues spoke publicly about them. Her daughter Alison felt one comment was particularly fitting of her mother. So many doctors hope to cure cancer. "She was one of the few people that actually got to do what she wanted to do with her life."

BIOLOGY AND THE ENVIRONMENT

MARIA SIBYLLA MERIAN

1647–1717

ENTOMOLOGY • GERMAN

MARIA SIBYLLA MERIAN LOVED BUGS LONG BEFORE SCIENTISTS had uncovered their mysteries, loved them at a time when few people were interested in those vile, disgusting things. Acquaintances assigned credit or blame for her unusual passion to her mother, who had looked at a collection of insects while Merian was still in the womb. Something about those pinned and polished bodies, shimmering powdery wings, and articulated legs instilled a fascination in the child growing inside her.

As a youngster, Merian kept a record of her beloved creatures (and their favorite hideouts) by learning how to draw them. Merian's stepfather was a painter and art dealer, and she learned from him how to mix pigments for watercolors by working the fine grains with a mortar and pestle, dropping the powder into water, and then sealing the solution with acacia tree sap, which helped the color bind to the page. To understand anatomical forms, Merian traced existing work, following the powerful zigzags of a grasshopper's leg and the creases of a snail shell as it spiraled outward.

At age thirteen, Merian started bringing her bugs into her home. She nurtured a little colony of silkworms, feeding them mulberry leaves or scraps of lettuce in a pinch. Merian took notes and painted the specimens as they mowed through their food, spun themselves into a "date pit" (a German expression for a cocoon), and burst open. She waited with giddy anticipation to see what would emerge. A wet moth? A cloud of flies? Nothing at all? Merian painted every stage.

Naturalists weren't paying much attention to insects during the time Merian studied them. Even the bugs' ability to reproduce was largely a mystery. When flies sprouted from rotten meat or dung, many believed they had spontaneously generated. There were even recipes that explained how, with a few simple ingredients, one could grow creatures like bees and scorpions. Want a worm? Mix one part dead flies and one part honey water on a copper plate. Warm the plate with smoldering ashes until . . . Voilà! Newly minted worms.

A well-known naturalist claimed that butterflies actually lived within caterpillar bodies—and that he could do a fancy trick with boiling water, vinegar, and wine to prove it. When previous illustrators drew the stages of a silkworm's metamorphosis, each form was filed separately, worms next to other worms and butterflies with butterflies. Their entire life cycle was hidden.

Merian saw each stage in a bug's life cycle as a continuous process, when few other naturalists were making connections between the worm and the butterfly. She situated her specimens in context, capturing them crawling on leaves with curling seams, flying over flowers' stretching tendrils, or circling around stems, at a time when most illustrators worked from display cases.

In 1679, Merian published her first major work on insects, a two-volume book of entomological illustrations focused on metamorphosis. With notes on food preferences and activities logged next to each image, Merian placed herself firmly within the tradition of naturalist observation.

As her career evolved, so did her personal life. In rapid succession, Merian split from her husband and moved from her native Germany to Holland with her mother and two daughters to join a religious sect. The group wasn't big into personal possessions, so for a while Merian's art went dark.

By 1691, the sect was flailing. It struggled to keep its European members healthy and support expeditions to Suriname, a

Dutch colony in South America, where the group hoped to set up a homestead. In one embarrassing incident, pirates robbed and stripped naked one of the sect's convoys, leaving them to arrive unclothed.

When the sect in Holland dissolved, Merian and her daughters relocated to Amsterdam. As hope for a new religious community abroad faded, Merian felt her own personal interest in Suriname swell. Over the years, she'd collected bugs on bridges, in backyards, in rural fields, and in meticulously manicured gardens. Friends boxed up their exotic finds and shipped them over for her to observe. After a lifetime studying the same specimens, Merian desperately wanted to go somewhere where she could discover more.

In 1699, at age fifty-two, Merian and her youngest daughter loaded up the art supplies and hopped on a ship for Suriname, financed by years of commissions and the sale of 255 paintings. The goal was to devote five years to exploring and illustrating the insects abroad.

In Suriname, a whole world of new specimens kept her plenty busy—and occasionally at risk. She braved the appealing fuzz of a red and white caterpillar, its poisonous barbs in disguise. But for Merian, the danger made the discovery even more interesting. She wrapped up the caterpillar and carried it home. Her expeditions moved incrementally further afield, and when she finally felt comfortable enough to go out into the rain forest, Merian followed a path forged by slaves, fresh-cut for specimen-collecting expeditions.

The bugs brought Merian to Suriname, but they also sent her home. Forced back to Europe three years early by malaria and the heat, Merian was still able to whip her two years abroad into her life's greatest work. *The Metamorphosis of the Insects of Suriname* was published in 1705, when she was fifty-eight. The book included sixty engravings that illustrated a creature's entire life cycle—just as she had done in her youth—with notes about its habits and environment. The vivid, writhing animals nearly crawled off the page.

The book brought Merian to a final transformation. One of the very first entomologists, Merian broke new ground in observing and documenting the stages of metamorphosis. By treating the insect life cycle as something worthy of rigorous study, she ushered in a new wave of scientists who followed her lead. Thirty years after *Insects of Suriname* was published, a French biologist developed the first classification system for bugs. Merian set the stage for one of the most significant moments in entomological history.

JEANNE VILLEPREUX-POWER

1794–1871

BIOLOGY • FRENCH

JEANNE VILLEPREUX-POWER SPENT A DECADE OBSERVING THE ocean's creatures before her work was swallowed by it. Villepreux-Power wasn't on the ship when it sank, but years of her scientific research plunged into the depths. The loss was substantial, but if anyone was capable of bobbing above the water, it was Villepreux-Power. She had already reinvented herself twice.

Growing up a cordwainer's daughter in tiny Juillac, France, Villepreux belonged in a larger arena. At eighteen, she left her hometown for a place grand enough to match her abilities and interests: Paris. Some say she walked there; others say she found a ride. Either way, the journey was fueled by determination.

In Paris, Villepreux landed a position as an assistant to a dressmaker, where she watched, worked, and experimented. In a few years she was able to prove her own considerable talents. In 1816, when Princess Caroline, the daughter of the king of the Two Sicilies, married Charles-Ferdinand de Bourbon, the nephew of King Louis XVIII, she was wearing a dress designed by Villepreux. The garment captivated Europe's upper ranks, and Villepreux—still in her twenties—found herself courted not only for her clothes but also for her hand.

Villepreux became a Power two years later, when she married the English merchant James Power based in Messina, Sicily. As she began her life on the island, Villepreux-Power realized the location offered her another opportunity for reinvention. Sicily was rich with varieties of flora and fauna unfamiliar to her. To learn more about her adopted environment, Villepreux-Power began teaching herself natural history while she embarked on a project to take inventory of the island's ecosystem. The goal was to catalog the plants, animals, and ocean life surrounding her waterfront abode.

In 1832, Villepreux-Power began studying a tiny relative of the octopus, called the paper nautilus. Its shell, which the nautilus uses to navigate the ocean waters, had been a mystery to scientists dating as far back as 300 BC, when Aristotle hypothesized that the creatures used their tentacles as both oars and sails in order to steer their brittle vessel like a boat. For centuries both the shell's utility and origin remained unknown, but in the nineteenth century the prevailing thinking was that it was an acquired home, like the ones hermit crabs procure. Villepreux-Power wasn't so sure.

The newly minted naturalist knew that there was only so much one could learn by plucking a creature from its briny environment. So in 1832 she invented a container that would facilitate her observations by keeping aquatic creatures alive in their own ecosystem—even while extracted from the ocean. She designed a glass case, the first recognizable aquarium. With it Villepreux-Power was able to watch her subjects long enough to discover that the shy paper nautilus doesn't scavenge her chambered shell at all; she *produces* it herself.

For the scientific community, both the stage Villepreux-Power designed for her experiments and the results gathered within it were big revelations. The British paleontologist Richard Owen (the man responsible for coining the word *dinosaur*) anointed Villepreux-Power the mother of *aquariophily* in 1858. The Zoological Society of London named the boxes "power cages" after their polymath inventor.

Eleven years after designing her first "power cage," Villepreux-Power continued to experiment with boxes she dipped in the sea. During that time she added wooden exoskeletons and anchors to a submersible model so that they could plunge deeper into the ocean. In her cages, Villepreux-Power watched starfish act out their private rituals of meal preparation and assessed the stomach content of mollusks.

Over her lifetime, Villepreux-Power gained membership to more than a dozen scientific academies throughout Europe—the Zoological Society of London and the Gioenian Academy of Natural Sciences in Catania, Italy, among them. After her death in 1871, *The North American Review* called her "one of the most eminent naturalists of the century," and the aquarium as "incalculable" to marine zoology. In 1997, Villepreux-Power ascended even higher when a large crater on Venus was named in her honor.

MARY ANNING

1799–1847

PALEONTOLOGY • BRITISH

BEFORE SHE WAS STRUCK BY LIGHTNING, MARY ANNING WAS A dull child. But after she was lifted from the grisly scene and sponged off (her babysitter and two friends dead and a horse-riding event ruined), the baby had changed. The once-placid infant had been zapped into a new state, forever after described as "lively and intelligent."

In a life filled with difficulties, electrocution was a rare (if bizarre) stroke of luck. Anning's family was poor. Of ten children, she and her brother were the only two who survived into adulthood. Her father was a carpenter who supplemented his meager income by hawking seaside souvenirs to tourists. The most sought-after trinket: fossils.

Anning's father pulled his specimens from limestone and shale cliffs of Lyme Regis, England. That ragged edge of their hometown ran along the sea. When a storm thundered in, large sheets of rock would tumble into the water, exposing sections of the area's history. Swooping in at just the right moment, Anning's father would find an assortment of shells and bones ripe for the picking.

Anning learned the trade from her father at age ten. After he died in 1810 from tuberculosis, Anning and her brother made the trips to the bluffs alone. Their haul was mostly shells and small fossils in the beginning. But in 1811, Anning's brother Joseph noticed a face emerging from the rock. Several weeks later, with a small hammer, Anning carefully cleared the sediment from around the curvature of the skull. The more work she put in, the more work there was to be done. The skull led to a spine and then a rib cage and legs. All in all, Anning traced around the bones of a beast some seventeen feet long with massive crocodile-like jaws. Two children discovered the world's first ichthyosaur fossil.

They sold the ichthyosaur (which means fish-lizard) to the lord of a nearby manor for twenty-three British pounds—several hundred dollars in today's currency. The ichthyosaur marked Anning's first substantial contribution to paleontology, but the fish-lizard was only the beginning.

Anning and her brother were not the first people to discover fossils in Lyme Regis. Locals had picked up strangely shaped bones here and there. Some believed that they were God's embellishments, and others thought the fossilized remnants might have come from the flood that lifted Noah's Ark. Anning's bones, however, told another story. By excavating fully articulated creatures from Lyme Regis's unstable rock, she had revealed specimens unlike anything anyone had ever seen.

With her dog acting as her sidekick after her brother lost interest, Anning surveyed the cliffs following storms and landslides, combing through the debris for specimens. The stones, shells, and bones she retrieved filled up a tiny roadside shop.

In 1823 Anning discovered a Plesiosaur (then referred to as a sea dragon) and five years later she delivered a Pterodactylus (called a flying dragon). Anning's ability to spot specimens, sort them, sketch them, and present them was unparalleled. She studied up on the ancient reptiles she found. Anning's moneyed patrons were routinely impressed by her breadth of knowledge.

Scientists profited greatly from her work, but because of her class and gender, the academic discussions sparked by her findings always excluded her. When Mary Anning's discoveries appeared in journals, her name was edited out. Anning's patrons arranged for a small stipend to fund her collecting, but the real profit—scientific acclaim—went to others.

Anning's accomplishments weren't respected in Lyme Regis,

either; neighbors considered her nothing more than a tourist attraction. To a young correspondent in London, Anning wrote, "I beg your pardon for distrusting your friendship. The world has used me so unkindly, I fear it has made me suspicious of everyone." She spent her life poor and largely alone. Her dog, Tray, was killed in a landslide.

The record of Anning's contributions has always teetered dangerously close to being covered over. In 1859, twelve years after Anning's death from breast cancer at age forty-seven, Charles Darwin published *On the Origin of Species.* The work was likely influenced by Anning's prehistoric discoveries. There were a few bright flashes of recognition along the way. In 1865, Charles Dickens wrote an article about Anning's life in *All the Year Round*, a journal he edited. It concluded with the line, "The carpenter's daughter has won a name for herself, and has deserved to win it."

ELLEN SWALLOW RICHARDS

1842–1911

CHEMISTRY • AMERICAN

BEFORE 1887, WATER QUALITY STANDARDS IN MASSACHUSETTS did not exist. Modern, city-run water treatment plants? Those weren't around, either. So on the contaminated drinking water roulette wheel, to take a sip of water in Cambridge, Massachusetts, in the late nineteenth century was to consume either industrial waste or municipal sewage. To push the area's drinking water to a safer state, Ellen Swallow Richards, an instructor at the Massachusetts Institute of Technology's newly founded laboratory of sanitary chemistry, supervised the collection and analysis of some twenty thousand water samples. Her experimental design both set the standard for similar studies and gave Richards a foundation to make assumptions about both the area's water quality and larger global drinking water conditions. Not a bad contribution from the first person in the United States to be both a professional chemist and a woman at the same time.

Richards believed that science could do an extraordinary amount to improve the public's daily lives. To tackle problems like water contamination, scientists and government are absolutely on the hook for ensuring that municipal resources are safe to use. But Richards also believed that by extending sanitation standards and basic science into the home, researchers would see vast improvements in the public's health. (The field of sanitation engineering sprouted in the late 1800s, thanks largely to Richards's work.) In addition to being one of the earliest voices in the field of ecology, she's also known for founding another major area of study: home economics.

A quick bit of context: When Richards entered MIT in

1870, she was the first woman ever admitted to the university. MIT let her attend tuition-free as a sort of insurance policy. Should anyone affiliated with the university complain about her, MIT could claim she wasn't really a student and "that her admission did not establish a precedent for the general admission of females." At the time, Richards was oblivious to the reasoning behind her status at the university. She later admitted, "Had I realized upon what basis I was taken, I would not have gone."

Richards earned both a bachelor's and a master's degree at Vassar and then a second bachelor's degree in chemistry from MIT. But when she started work on her PhD, MIT put the kibosh on her progress. The university simply wasn't ready to bestow the honor on a woman.

Richards did not comport herself as an exceptional case, or as a member of a rare clan of women who could hack it with the men. If she was going to get a good education (or most of one), she would make sure to extend the opportunities to others wanting to do the same. However, MIT still wasn't officially open to women. With funding and initiative provided by the Woman's Education Association of Boston, Richards spearheaded the creation of a parallel science program for women on the MIT campus. Opened in 1876, the Women's Laboratory at MIT was a place for budding scientists to conduct research and take classes. The lab was two rooms flanked by big windows that showed off the spectacle: women studying industrial chemistry, mineralogy, and physiology. In a report on the program, she wrote, "I have felt the greatest satisfaction in opening the treasures of our store-house."

Her influence quickly spread beyond her charges in the lab. Richards wrote letters to women who enrolled in correspondence courses as a part of an effort initiated by the Society to

Encourage Studies at Home. The idea was that Richards would teach her distant pupils science, but her advice was soon requested for a myriad of problems. Conditions at home were bleak. In their letters women said they were overworked, and ill health was a common theme.

The concerns spurred Richards to act. She wanted to incorporate science-based advice into her prescriptions for improving life in the home, and so she began talking to her pen pals about eating a more balanced diet, preparing healthy foods, exercising regularly, and wearing comfortable clothes (corsets were still in fashion).

In ironic testament to its success, the Women's Laboratory shut down in 1883, when women were finally admitted to MIT's standard set of science classes. Shortly thereafter, Richards began her groundbreaking work on water sanitation and started a position as a chemist and a water analyst for the Massachusetts Board of Health. Concurrently, Richards formulated a plan to deliver science to women.

In 1890, Richards's efforts to address the dearth of information about nutritious, inexpensive, and safe food preparation resulted in the opening of a kitchen that both served food and provided a hands-on education for the public. Four years later, the kitchen started supplying nutritious meals to schoolchildren. (The program predated Michelle Obama's healthy school lunch initiative by 116 years.)

Richards also advocated for domestic science to be taught in public schools. Her efforts rolled out slowly, but gradually they became a movement. Richards published books, gave speeches, and in 1908 the American Home Economics Association was founded with Richards as its president. Home economics became a major conduit to bringing women into science at the university level.

Richards had the extraordinary vision to see how science's influence could extend in all directions, from sanitation to conservation to education, home, health, and happiness. All it took was a little knowledge and, well, twenty thousand water samples.

ALICE HAMILTON

1869–1970

BACTERIOLOGY • AMERICAN

ALICE HAMILTON'S PROFESSIONAL SUCCESSES—OF WHICH there were many—fell at the intersection of science and social issues. Although she earned a degree in medicine from the University of Michigan, gaining further training in bacteriology and pathology at the University of Leipzig and the University of Munich, she didn't think herself capable of becoming anything more than a "fourth-rate bacteriologist." But what she lacked in bravado, she made up for in her dedication to problems both "human and practical": typhoid outbreaks, lead poisoning, and the widespread horror of occupational disease.

One of her earliest gigs was to help a neighbor nail people peddling cocaine to children. The time was roughly *Newsies*-era, but in Chicago—the very beginning of the twentieth century. The city was having a problem with drugstore employees offering children a sample of "happy dust" on their way home from school. One boy reported that the powder made him feel "as if I was going up in a flying machine," and another claimed it was "as if I was a millionaire and could do anything I pleased." When they wanted more, the children had to pay. Some of the most desperate resorted to smashing drugstore windows, robbing people, and intimidating clerks for cash. Social reformers swooped in to try to get the dealers off the streets and away from the children. Hamilton was called in as reinforcement.

In order to testify about cocaine in court, Hamilton learned how to verify that the powder confiscated was indeed "happy dust." The lab results, though, were imprecise. The test would ring positive whether the powder was cocaine or the synthetic alpha- and beta-eucaine, a cocaine derivative. Because the latter substance was legal, the test became a boon for defense attorneys, who gleefully took advantage of the loophole to argue away the charges. But Hamilton had another idea for how she might make the accusations stick. Cocaine, when dabbed on the eyeball, will dilate the eye. When the same is done with alpha- and beta-eucaine, the eye's appearance remains unchanged. At first she performed the trick on rabbits, but the method didn't engender any kindness from the jury, who quickly turned their sympathy toward the bunnies. "So I tested the powders on myself," she admitted. "For I knew it would not injure my eyes and other people were quite understandably reluctant to take the risk. I used to go around the laboratory with one wide and one narrow pupil till everyone was so used to it that they took no notice." After a year, she and a few others succeeded in broadening the law's coverage to include eucaines.

Hamilton was never afraid to get her hands dirty. Take her approach to the epidemic of typhoid fever that broke out in Chicago in 1902. The area most severely affected was the one directly around Hamilton's residence. With her training in pathology and bacteriology, a friend figured that if Hamilton could root out the outbreak's cause, the city could formulate a solution.

First Hamilton investigated the water and milk supplies, but neither explained why the nineteenth ward specifically was hit so hard. Next, she surveyed the neighborhood, hoping that visual clues might lead her to the answer. "As I prowled about the streets and the ramshackle wooden tenement houses, I saw the outdoor privies (forbidden by law but flourishing nevertheless), some of them in backyards below the level of the street and overflowing in heavy rains," she explained. "The wretched water closets indoors, one for four or more families, filthy and with the plumbing out of order because nobody was responsible for cleaning or repairs; and swarms of flies everywhere." And there it was: the flies.

One way typhoid is transferred is by exposure to contaminated sewage. Perhaps, Hamilton thought, the flies were gobbling up the diseased human waste and then landing on uncovered food and milk, spreading the disease.

Hamilton put her theory to the test by collecting the pests from kitchens and both indoor and outdoor bathrooms. Sure enough, the flies were carriers of typhoid bacillus. Hamilton's findings squared well with previous observations during the Spanish-American War, and they also explained why the better-off—those with reliable plumbing and screened-in eating areas—weren't having the same problems. Presented to the Chicago Medical Society, Hamilton's paper won lots of attention, prompting a total reorganization of the health department, including the addition of an expert dedicated to inspecting tenement housing.

Although the outcome was positive, Hamilton's tidy explanation was incorrect. As she later found out, the real cause of the typhoid outbreak was one actively covered up by the Board of Health; a sewage spill had contaminated the water supply in the nineteenth ward for three straight days. "For years," admitted Hamilton, "although I did my best to lay the ghosts of those flies, they haunted me and mortified me, compelling me again and again to explain to deeply impressed audiences that the dramatic story . . . had little foundation in fact."

But uncovering these truths—no matter how buried in muck—was what made Hamilton so exceptionally effective at assessing unsafe environments. Remarkably skilled at gathering information from sources who didn't want to provide it, Hamilton was able to make tremendous headway in unsafe industries by asking workers, *Why do you keep a job that's clearly killing you?* She conducted interviews in the home, where she figured workers would be comfortable and forthcoming. During one such visit, Hamilton asked a man suffering from lead poisoning why he continued showing up to work. House payments and a family stopped him from quitting, he said. The plants often preferred to hire married men. Hamilton suspected the choice was calculated; out of an obligation to support their families, workers were less likely to quit, even when lead was causing colic, convulsions, and weight loss.

In 1910, Hamilton's focus shifted to health in the workplace full-time, when she was asked to serve as the managing director of the Occupational Disease Commission in Illinois—the first commission of its kind in the country. The task was to survey the state's "poisonous occupations," to figure out what kinds of plants were exposing workers to harmful substances like carbon monoxide, arsenic, and turpentine and to assess how many plants existed. The team split up by noxious substance. Hamilton took lead. At the start of the project, the government knew neither what industries manufactured with lead nor how pervasive its ill effects were.

Hamilton dug in, starting with the most obvious lead-using industries and hoping those inquiries would draw her nearer to ones invisible. She and her team approached the project with a detective's scrutiny by visiting plants, interviewing doctors and industry leaders, and combing through hospital records for telltale signs of patients with lead poisoning. Her inquiries shook out a long list of processes that required lead, including freight-car seals, coffin trim, glass polishing, and "tin foil" cigar wrappers ("tin foil" turned out to be a misnomer). Hamilton found buildings dilapidated and improperly vented, with lead dust clouding the air even around workers who didn't produce it. In one plant, an astonishing 40 percent of employees had gone to the hospital for becoming "leaded."

By 1919, Hamilton was the foremost expert on industrial health in the United States. So when Harvard decided to expand their curriculum to include public health, Hamilton noted, "I was about the only candidate available." Hired as an assistant professor of industrial medicine, Hamilton became Harvard Medical School's first female faculty member, predating the arrival of female medical students by twenty-six years. (Her appointment came with three stipulations: she was not to set foot in the Harvard Club, not to claim faculty-reserved tickets for football games, and not permitted to participate in commencement ceremonies.) The appointment caused quite a sensation, but Hamilton recalled the welcome as warm.

While working half-time at Harvard and half-time doing fieldwork, Hamilton investigated carbon monoxide poisoning for the US Department of Labor. She also looked into the virulent effects of aniline dyes, mercury, volatile solvents, and other toxic products. Her reputation spread. General Electric tapped her as a medical consultant, the President's Research Committee on Social Trends appointed her as a member, and she was brought onto the Health Committee of the League of Nations and the Public Health Service of Soviet Russia to consult on industrial hygiene.

In her memoir, *Exploring the Dangerous Trades*, Hamilton expressed pleasure in leading messy, deleterious, misinformed industries to a healthier future. "No young doctor nowadays can hope for work as exciting and rewarding. Everything I discovered was new and most of it was really valuable." The industries to which Hamilton applied her unique expertise underwent dramatic change as a result. After her yearlong lead study in Illinois, the state passed a law that compensated workers for harmful exposure to noxious gases, dusts, and fumes. The law set off a systemic change. Because employers began to insure against such health-related claims, insurance companies responded by pushing for workplace reforms. By 1937, most of the states carrying the country's largest industrial burdens had adopted legislative requirements that workers had to be paid for being poisoned.

By knocking on the doors of a city's poorest residences and applying pathology to the problems she observed, Hamilton was able to record solid evidence of occupational illness. Her pioneering determination paved the way for real social change.

ALICE EVANS

1881–1975

BACTERIOLOGY • AMERICAN

ALICE EVANS DIDN'T FIND IT CHALLENGING TO RESPOND TO her critics. If her peers had specific questions about her experimental design or her results—that two strains of bacteria considered separate were actually related—she had no problem filling in the details. But the response, she recalled, wasn't really about the evidence at all. "The reaction to my paper was almost universal skepticism, usually expressed by the remark that if these organisms were closely related, some other bacteriologist would have noted it."

The organisms in question were *Bacillus abortus* and *Micrococcus melitensis*. The former is a wicked bacteria that affects cattle, causing weight loss, infertility, lower milk production, and for cows gestating, spontaneous abortion. In the early 1900s, infection was expensive for farmers and unpleasant for cattle. Similarly nasty is *Micrococcus melitensis*, a highly contagious bacteria among animals that also affects humans, causing rolling fevers, chills, and aches that, at the turn of the century, would get into human bodies and stick around for decades. Before Evans, the two bacteria were considered entirely separate. After Evans, the strains were shown to be not only related, but capable of jumping from animal to human. In 1917, most scientists found Evans's ideas just too radical to believe.

Back to her critics' question: Why had someone else (read: a man) not seen the similarities before? Evans explained that there was an initial mistake made in the bacteria's classification. The discoverer of the one responsible for illness in humans described it as a sphere, so he filed it into a genus with other spheres. The scientist who discovered the bacteria that affects cattle considered it rod-shaped, and so he classified it as such.

The epidemiologist who was among the first to spot *Bacillus abortus* in cow's milk, and who was fervently opposed to the idea that it might harm the milk's drinkers, was so outraged by Evans's claims that he refused to chair a committee that she sat on. Farmers accused her of conspiring with equipment suppliers to push for pasteurization. Didn't she know it was common knowledge: Milk was nutritious. Wouldn't fresher milk be better? In 1918 Evans published her results in the *Journal of Infectious Diseases* and waited for public acceptance.

Evans came to science as an adult because she got a small taste of experimentation and realized that nothing else she'd ever done had felt as galvanizing. She started her career as a schoolteacher in rural Pennsylvania, but after four years of blackboards and pencil sharpeners, she was flat-out bored.

In 1905, she took advantage of a tuition-free nature study program offered to rural teachers. As soon as Evans got a taste of biology she lost interest in teaching. Anxious to learn more, Evans stopped caring about finishing her two-year certification.

Instead, she worked toward a bachelor's degree from Cornell and then was awarded a scholarship to the College of Agriculture at the University of Wisconsin to earn her master's. Her professors at Wisconsin were supportive, urging her to enhance her chemistry skills and offering her a glimpse at heavyweight-level techniques. (Three years after she graduated, a former professor would discover vitamin A.)

When faced with the decision to get a PhD or get her slides dirty, Evans realized she was eager to enter the field. The US Department of Agriculture first brought her on as a bacteriologist in 1910 to study cheese—specifically to find ways to make it tastier. Three years later, when the USDA set up its new Dairy Division lab in Washington, DC, she followed. Four years after that, Evans examined two seemingly different bacteria on a glass slide, and called them siblings.

The same year she published her findings, Evans moved over to the Hygienic Laboratory of the Public Health Service. Influenza, infantile paralysis, sleeping sickness, she tackled them all. Several years after her discovery, support for her controversial research began to roll in. A scientist from San Francisco confirmed her results, and researchers around the world were coming to the same conclusion.

When the controversy finally disappeared, *Bacillus abortus* and *Micrococcus melitensis* were joined in a new genus: *Brucellosis*, which is what we call both related strains today. However, filing them away didn't mean Evans was done with them. In 1922, Evans was infected with the disease that she'd made her name studying. She suffered from the effects of *Brucellosis* on and off for more than twenty years.

Evans's research changed both minds and regulations. In the 1920s, standards were enacted that governed the state a barn needed to be in order to be used for milking. (Hint: very clean.) In the 1930s, pasteurization of milk became mandatory, due in no small part to Evans's work. She was lauded for her scientific contributions with honorary degrees, places on delegations, and appointments to scientific societies. Finally, universal skepticism was met with universal acclaim.

TILLY EDINGER

1897–1967

PALEONEUROLOGY • GERMAN

LONG AFTER MOST SCIENTISTS WITH JEWISH ANCESTRY HAD been fired from their positions under Germany's Nazi regime, Tilly Edinger tended to her collection of fossils as she always had, organizing and acquiring fish, mammals, reptiles, and amphibians. She examined fossilized skulls and considered what these bones might tell researchers about ancient brains. With a streak of paleontological humor, Edinger liked to say that she hung on to her position at the Senckenberg Museum of Natural History in Frankfurt like an ammonite in the Holocene.

The museum successfully avoided releasing Edinger for years because, well, could it really fire someone unpaid? Besides, the institution was privately owned, not publicly held. Edinger had the museum's staff on her side. She recalled a colleague "fighting like a hero to keep me in the house." Even as the dangers of staying in Germany became more apparent, Edinger was reluctant to leave Frankfurt and her family's 380-year-old history in the city. While Edinger held out for better times, she prepared for the worst. On her person, Edinger carried a fatal dose of a sedative that she vowed to take should she ever be hauled off to a concentration camp.

With some thirty thousand Jews arrested and one hundred killed during Kristallnacht—the Night of Broken Glass—Edinger changed her stance on staying. Widespread violence and the threat of more forced the museum to cleave Edinger from her work. She was barred from entering the museum, and the contents of her office were shipped to her home without comment. Edinger was one of the very last Jewish scientists in Germany as the time to hold on to employment. The dismissal was devastating, and her refusal to leave Germany put her in danger. Even so, Edinger had some sense that "the fossil vertebrates will save me." They had, after all, been her life for nearly twenty years.

Edinger came to paleontology in college, after finding her classes in zoology unfulfilling. Like her father, who was a well-known neurologist, Edinger was also fascinated by the brain. Her specialty was the prehistoric variety, which she could study by examining the inside of ancient skulls. For her PhD dissertation in 1921, Edinger launched an investigation into the brain of the Nothosaurus, a massive, extinct marine reptile.

Within her first decade of carefully studying skulls at the Senckenberg Museum of Natural History, Edinger launched a new field: paleoneurology. The field's founding document was a 250-page review article that Edinger produced by gathering disparate shreds of published work on fossil brains, organizing the research by specimen, and then summarizing the conclusions that could be drawn from so many previously isolated specimens. She laid out a detailed history of the field, a thorough explanation of what was presently known, and then identified the big questions that still remained. In the phylogenic section, she all but dismantled another scholar's largely accepted laws of brain growth. Edinger's work was widely recognized and admired throughout Europe. In the middle of World War II, the review would serve as her leverage to leave Germany.

Research positions at American universities saved numerous scientists from the horrors of the Holocaust. Edinger started lobbying late, but the scientific community mobilized around her. The American bacteriologist Alice Hamilton, a family friend, begged Harvard to hire Edinger. Others sent letters to the US government making a case for her entry: "She is a research scientist of the first rank and is favorably known as such all over the world," wrote the American paleontologist George Gaylord Simpson. She had founded paleoneurology, "a study of outstanding value and importance," he pressed. As she waited for her number to come up in the United States, Edinger escaped to London, where she spent a year translating German texts as a part of the Emergency Association of German Scientists in Exile program.

Edinger was approved for relocation stateside in 1940, and when she arrived, the Museum of Comparative Zoology at Harvard immediately picked her up as a research associate. With scientists singing in the prep room and whistling in the halls, Harvard provided a soft landing. Finally settled, it was time for Edinger to get back to the business of paleoneurology.

Looking at the inside of an extinct animal's skull for clues could reveal an extraordinary amount about the size and structure of parts in an ancient brain. Map these structures over time and connect them to brain function, and the data could reveal a fascinating history of a species. It was through this method that Edinger proposed that whales used to rely on their sense of smell a lot more than they do today. How could she tell? Edinger compared casts of the inside of whales' skulls, both ancient and current. In more ancient animals, there was more space for the amygdala, the section of the brain that handles smell. Edinger observed how the skull structure adjusted over time as the brain's olfactory lobes shrank.

A complicating factor in reading these casts is the so-called braincase, the specific portion of the skull in which the brain rests. Nailing down the particulars of one creature's brain padding says nothing about the next. In fish, reptiles, and amphibians, the meningeal layer and vascular tissues are thick; in birds and mammals, they're thin. To make an educated guess about an extinct animal's skull-to-brain ratio, Edinger leaned on analogues of ancient species, alive today.

Working from an inquiry started by her father back in Germany, Edinger suggested an area of research that could be tackled in the United States. With the nation's abundant supply of equine records, she thought there would be easy access to a collection of casts documenting the structures of a horse's brain over time. Upon arriving at Harvard, a colleague challenged Edinger to take up the horse study herself. Edinger had a terrible time tracking down the materials she needed. It took her a decade to complete the monograph on horses, but her conclusions were crucial: in 1948 Edinger reported that brains and bodies of the same species did not evolve in unison, and different mammals underwent evolutionary changes at different times.

Through all her struggles—political, work, or personal—Edinger had a bright sense of humor, even in the midst of long academic disputes. She and Princeton-based Glenn Jepsen argued for some time over a fossil skull that could belong either to a bat or to a miacid, depending on whose side you were on. Her good nature was reciprocated by Jepsen, who wrote a poem parodying her position:

THE TILLYBAT
A curious beast is the Tillybat
It surely seems odd and quite silly that
With a brain shape so batty,
We'd find glenoids so catty!
You see why we call it a dilly, Pat?
"The midbrain is hilly,—
and further," says Tilly,
"Look here quick and see

Those colliculi!
It had to squeak, not mew,—
it never walked, it flew!
Jep, don't be so placid,
It's not a miacid!"

Jepsen surely gained some points with Edinger for creativity; however, she maintained her original opinion.

The last years of Edinger's life were dedicated to translating her previously published 250-page tome into English and updating it with new information. When she'd researched and written the first edition at the Senckenberg Museum, she'd done the entire thing without staff support. As she returned to the project, Edinger often switched off her hearing aid to tune out her colleagues, finding comfort in the silence once again.

In 1964, Edinger was given an honorary degree by the University of Frankfurt. Having been forced out of her town and her country, she was touched by the gesture. It had been more than twenty-five years since she'd left. The degree was a tangible sign of change.

RACHEL CARSON

1907–1964

MARINE BIOLOGY • AMERICAN

WHEN WE TALK ABOUT RACHEL CARSON, WE TALK ABOUT *SIlent Spring.* First serialized in *The New Yorker* before being published in book form in 1962, *Silent Spring* chronicled the devastating effects of the overuse of pesticides. The book was startling for its rigorous scientific assessment of how, by spraying for one issue—to get rid of a bug or a weed—without considering how the chemicals would impact everything else, people were often doing more harm than good. It was a beautifully written treatise of horrors aimed at a general audience.

Silent Spring jump-started the environmental movement and provided the public with a target: the multimillion-dollar chemical industry. In turn, the chemical industry reacted by launching a quarter-million-dollar smear campaign against Carson. She was called hysterical, labeled a spinster, and accused of letting harmless insects terrify her. Whenever Carson or the book set off an outrage, the chemical industry fanned the flames. As a result, in the early 1960s, Carson was at the center of a very public battle between those hoping to preserve nature and those wanting to control it. Fortunately, it was a fight Carson had been preparing for her entire life.

Carson had loved the outdoors ever since she was a child. The birds and plants she observed around her family's rural property and along Pennsylvania's Allegheny River had sparked her imagination for as long as she could remember. Over the years, she found a fossilized fish, hopping birds, and native plants. Inspired by her expeditions, at age eight Carson wrote a book about a pair of wrens looking for shelter she titled *Little Brown House.* Her well-written tales and her persistence in pitching them allowed Carson to join an underage literary elite of young contributors with published works in the now-defunct children's magazine *St. Nicholas.* (William Faulkner, F. Scott Fitzgerald, E. E. Cummings, and E. B. White also saw their bylines in the magazine as children.) Carson was fond of noting that she had become a professional writer at age eleven.

When Carson entered the Pennsylvania College for Women on a senatorial district scholarship, earning an English degree as preparation to become a writer was a natural choice. However, during her undergraduate years, it was biology that she found most thrilling. That fossilized fish? Biology gave her the tools to learn what had happened to it.

After earning a master's in zoology from Johns Hopkins, Carson found part-time work at the US Bureau of Fisheries. Though she'd chosen science over prose, her former specialty proved useful in her new occupation. Carson's first assignment for the bureau was to write a fifty-two-episode radio program called *Romance Under the Waters.*

"I had given up writing forever, I thought. It never occurred to me that I was merely getting something to write about." Her bosses at the bureau were thrilled with her work; however, the appreciation did not translate monetarily. To supplement her low pay, Carson churned out articles covering conservation issues for the *Baltimore Sun* on a freelance basis.

Although Carson climbed the ranks at the bureau to eventually become an aquatic biologist, her duties never included actual scientific work. Instead she was asked to do things like edit her colleagues' scientific reports and package a study's results into public brochures.

What she learned at the bureau during the day proved useful to her freelance career in the evenings. In 1937, Carson published a story in *The Atlantic* that examined the sea from the perspective of the animals and plants within it. The article's descriptions of aquatic creatures—and even of their deaths—were mesmerizing. "Every living thing of the ocean, plant and animal alike, returns to the water at the end of its own lifespan the materials that had been temporarily assembled to form its body. So there descends into the depths a gentle, never-ending rain of the disintegrating particles of what once were living creatures of the sunlit surface waters, or of those twilight regions beneath."

The article led to the publication of her first book, *Under the Sea-Wind*. Although it would be Carson's favorite, the volume was a commercial failure, selling only two thousand copies. Carson needed a couple of years to recover from the blow, but both driven and strapped for cash, she pushed forward. Carson wrote another book. When *The Sea Around Us* arrived in 1951, it won the National Book Award for nonfiction and solidified Carson's position as a literary heavyweight. To this day, it's credited as being one of the most successful books ever written about nature.

She found the public interest in nature extremely heartening. "We live in a scientific age; yet we assume that knowledge of science is the prerogative of only a small number of human beings, isolated and priest-like in their laboratories. This is not true," Carson said in a speech. "The materials of science are the materials of life itself. Science is part of the reality of living; it is the what, the how, and the why of everything in our experience."

By the time she turned her focus to pesticides in *Silent Spring*, Carson had the public's attention. The book's main target was DDT, or dichlorodiphenyltrichloroethane. DDT was the first modern, lab-made insecticide. Credited with curbing malaria and typhus in World War II, DDT was viewed as a panacea—something that fell under the umbrella of, as the DuPont company famously put it, "Better Things for Better Living . . .

Through Chemistry." In *Silent Spring,* Carson worried about its rapid and near-universal acceptance. "Almost immediately DDT was hailed as a means of stamping out insect-borne disease and winning the farmers' war against crop destroyers overnight."

So new and revolutionary was the poison and its ability to control pests, Carson argued, that proper precautions in understanding the greater effects of its application were not being taken. Using DDT was like flicking down one domino but ignoring the long line of others tumbling in succession behind it.

Now, Carson believed in science; her entire career was built upon her devotion to it. But by looking at pesticides from only one angle, she argued—as chemical companies were wont to do for their bottom line—they were being irresponsible. Carson laid out her case with scientific studies and observations from the field: twenty-seven species of dead fish in the Colorado River, a greenhouse worker with paralysis, accidentally poisoned livestock.

Alarmed by the information in *Silent Spring,* a US Senate subcommittee called Carson in to speak about her research, federal and state organizations started investigating the effects of DDT and other pesticides, and grassroots initiatives began to organize.

Silent Spring was tremendously influential. Three major events in 1970 were inspired by Carson. The National Environmental Policy Act promoted "efforts which will prevent or eliminate damage to the environment and biosphere and stimulate the health and welfare of man." A senator from Wisconsin later called it "the most important piece of environmental legislation in our history." In April of that year, the United States had its first Earth Day, and then the Environmental Protection Agency was formed. In a timeline of the EPA's history, *Silent Spring* is the first reference, the official germination of the agency. Carson wouldn't be around to see the changes called for by both the government and individuals as a result of her book. Breast cancer took her too swiftly, just two years after *Silent Spring*'s publication. But her book succeeded in bringing about change. Carson's resounding voice is embedded in the foundation of modern environmentalism.

RUTH PATRICK

1907–2013

BIOLOGY • AMERICAN

ONE AFTERNOON IN THE SUMMER OF 1959, RUTH PATRICK AND a colleague were traveling down a river in Ireland when suddenly they found their rowboat bow-to-bow with a British naval vessel. Over the loudspeaker, the bigger ship commanded that the pair "Come here at once." Annoyed, Patrick replied, "I will when I finish my business." It was important work, watching a cork float down the Lough Foyle. The cork's movements were essential to understanding the river's currents. Stopping would mean losing sight of the cork and the day's work. But the naval vessel insisted. "Come here at once or we'll shoot!"

The navy apparently mistook the cork for a snorkel and the snorkel as a sign of a bigger threat. After several hours of questioning, Patrick's employer stepped in to explain. Yes, she was measuring the current in preparation for a new DuPont chemical plant to be built in the area. An American ambassador later joked with her at a dinner event, "So you're the lady who was going to blow up the Queen's Navy! That story has gotten all over London!"

Ruth Patrick's influence by that point had nearly spanned the globe. Patrick was the first scientist to show how the health of rivers could be measured by looking at a body of water's tiniest organisms—single-celled algae called diatoms. "You see, diatoms are like detectives," she explained. A diatom's cell walls are made of silica, which takes in environmental pollutants. Almost proudly, Patrick told a local PBS affiliate she was able to detect radionuclides from Chernobyl that had been picked up by diatoms. These simple little organisms could reveal a body of water's history.

And sometimes they did. Patrick's investigation into the Great Salt Lake in the 1930s gave some solid clues about its origin. She found freshwater diatoms layered into the lake's deposits—up to the point when a change occurred. With no sand or chlorides, Patrick ruled out a tidal wave as the cause of a sudden shift from freshwater to salt.

Patrick realized that a river's overall biodiversity could tell scientists a lot about the body of water's health. Problems with contamination appeared in locations where relatively few organisms lived. If the assessment that thriving communities equal good ecosystems seems obvious now, it's because Patrick pioneered the idea. In 1954 she even invented a device that would take better water samples, called a diatometer.

Diatom sampling satisfied Patrick's adventurous spirit. At age seventy-six she wagered that she'd waded into some nine hundred rivers, spread out over every continent but Africa. That same year, she confronted 102-degree temperatures mucking around the Flint River in Georgia. Patrick pulled on her waders to gather river samples all the way into her nineties. When she couldn't go out anymore, she moved her work to the lab, coming in every day to analyze her "lovely" diatoms. A run-in with an aggressive foreign naval operation was just one tiny story in 105 years of them.

As a child growing up in Kansas City, Missouri, Patrick recalled eagerly that she "collected everything: worms, mushrooms, plants, rocks. I remember the feeling I got when my father would roll back the top of his big desk in the library and roll out the microscope. . . . It was miraculous, looking through a window at a whole other world."

She held that interest close as she took the study of botany from a bachelor's degree to a PhD, attending Coker College and the University of Virginia, respectively. Because the Academy of Natural Sciences in Philadelphia had America's best collection of diatoms, Patrick started working with the institution in graduate school. She had a long career ahead of her, and when Patrick thought about those early days, she remembered being "a little peon." After receiving her PhD in 1934, Patrick stayed on at the academy as a volunteer curator in the Microscopy Department. She took the existing diatom collection and grew it to become one of the most extensive in the world. It wasn't until 1945, though, that the organization came to its senses and started paying Patrick for her work.

Patrick dedicated her life to studying pollution in order to curb it. She took a unique path, choosing to work with big industrial clients like DuPont to lessen their negative environmental impact instead of shouting at them. "You can't have society without industry," she told a reporter in 1984. "But on the other hand, industry has to realize that it is a responsible group."

Patrick influenced the way people thought not only about rivers and lakes, but also about drinking water, which she showed was being contaminated regionally by fertilizer runoff and septic tank seepage.

Her opinion was sought after by a line of presidents. Lyndon B. Johnson asked for her input on water pollution, and Ronald Reagan consulted her about acid rain. For Patrick's work in industry, at universities, and at the Academy of Natural Sciences, Bill Clinton awarded her a National Medal of Science in 1996.

On the eve of Patrick's hundredth birthday, a reporter brought up a comment from a prominent environmental scientist regarding her legacy. "I try not to think about it," Patrick said. She still had several more years of important work to do.

GENETICS AND DEVELOPMENT

NETTIE STEVENS

1861–1912

GENETICS • AMERICAN

ARISTOTLE'S ADVICE FOR OLDER MEN LOOKING TO PRODUCE A male heir: have sex in the summer. The hotter the better. The ancient Greek philosopher believed that if enough heat was generated during conception, the baby would be a boy. A lack of fire in the sack would mean letting a woman's natural internal frigidity win; the baby would be a girl. So when a man wasn't able to generate enough passion or fire, one could fake it by relying on the outdoor temperature.

The idea that environmental factors determine a baby's sex persisted into the twentieth century (although other elements like nutrition were eventually added to the mix, too). After thousands of years of theorizing, in 1905 Nettie Stevens finally helped set the science straight. Chromosomes—not heat or diet or the side of the bed you roll out of—determine a baby's sex at the time of conception. And she had data from mealworms to prove it.

So ingrained was the idea of an outside factor nudging the fertilized egg one way or the other that it took years for Stevens's contemporaries to come around. When they finally did, it was too late for her to receive the recognition she deserved. Stevens died of breast cancer eleven years after her career started. The credit for the discovery of sex determination often defaults to Thomas H. Morgan, the geneticist responsible for discoveries pertaining to the role of chromosomes in heredity. When Stevens published her paper about sex determination, Morgan, an advisor of hers, stuck to the prevailing theory that sex determination had something to do with the environment, at least in the beginning.

Stevens was no stranger to playing the long game. Born the daughter of a carpenter and lacking financial support, Stevens bootstrapped her way through college. Until age thirty-five, Stevens hopped back and forth between taking classes and teaching them, saving all her pennies for a greater opportunity.

In the 1890s, Stanford University was brand-new, still called Leland Stanford Junior University. A major advertising campaign on the East Coast sought to entice students across the country to come to the Bay Area. Stanford was both cheaper than the University of California schools and it offered a unique open course structure. Students could take any classes that interested them, regardless of specialty. In 1896, Stevens, who was born in Vermont, enrolled at Stanford.

Always interested in science, Stevens was finally able to pursue biology. She spent much of her lab time corralling protozoa. As an undergraduate she discovered a new species, slotting it into a new genus of ciliates. Stevens's research was way out in front of her classmates'. She peered into the structures of the ciliates' cells, publishing a description of distinct chromosomes in protozoa for the first time.

Stanford provided a solid foundation in biology, but Stevens was a rising talent, drawn to the cutting-edge research happening back east at Bryn Mawr. At the turn of the century, the small Pennsylvania college was a hub for genetics talent. Morgan chaired Bryn Mawr's biology department, succeeding Edmund Beecher Wilson. With a brilliant grasp of chromosomal research, Stevens gelled with the department head and his research immediately. She also began pulling in fellowships, which sent her to Germany and Italy to further her cytology studies. In 1903, at the age of forty-one, Stevens received her PhD from Bryn Mawr.

The conclusion of her schooling brought Stevens back to a familiar issue: funding. Hoping to continue her research within the genetics community at Bryn Mawr, Stevens appealed to the Carnegie Institution for financial support. Both Morgan and Wilson sent in letters of recommendation. "I beg to urge as strongly as possible her appointment," wrote Morgan. "Of the graduate students that I have had during the last twelve years I have had no one that was as capable and independent in research work. . . . She has an independent and original mind and does thoroughly whatever she undertakes."

It worked, and Stevens was able to apply her energy to figuring out what determines sex. The study involved plucking the tiny gonads from mealworms, beetles, and butterflies and fixing them in a solution. Like trapping mosquitoes in amber, Stevens then secured the preserved sex organs in paraffin blocks in order to slice them into thin pieces without crushing the structures. Each slice was attached to a slide, dyed, and scrutinized beneath a microscope. When it was done just right, Stevens could see a whole line of chromosomes laid out before her.

Over time the biologist saw a pattern emerge in their structure. Male reproductive cells contained both X and Y chromosomes. Female cells contained only X. Harking back to Mendelian genetics and inherited traits, Stevens concluded that it was the combination of chromosomes at conception that determined a baby's sex. In 1905, Stevens published her theory in two parts. That same year, Wilson, another former advisor, arrived at the same conclusion mostly independently. His paper was more timid about upending two-thousand-plus years of thinking. Stevens was ready to be bold. "Here . . . it is perfectly clear that an egg fertilized by a spermatozoon containing the larger heterochromosome develops into a female." She had waited long enough.

HILDE MANGOLD

1898–1924

EXPERIMENTAL EMBRYOLOGY • GERMAN

WITH A TINY SET OF TOOLS, HILDE MANGOLD SLICED AMPHIBian embryos under a low-power microscope. The actual incisions were made with an ultrafine needle made of glass, its sharp end pulled to the tiniest point by the flames of a micro gas burner. (Micro because the burner was fitted with a capillary-sized flame-directing tube.) When the embryos needed to be flipped, Mangold nudged them with a loop made out of baby hair unwittingly donated by her advisor's child. The ends of the hair were brought together, shoved into another slim glass tube, and then secured with a bit of wax. Mangold carefully portioned off a specific chunk from one species' embryo and implanted it in a specific spot on the other species' embryo. The joined embryos were returned to pond water to grow. Removed from their protective membrane and exposed to the water's bacteria, most died.

Breeding season started in April, and it took Mangold two years to nurture six embryos to a reportable outcome. Mangold had found the embryos' so-called "organizer," the chunk of cells that grows the neural tube (basically the central nervous system and spine starter in fertilized eggs), and shown how it directs the growth of tissues and organs, bringing an animal to life. Implant any old group of cells into another embryo and those cells will grow into the normal flesh of the host amphibian. But scoop the organizer cells out of one embryo and add them to another—being careful not to disrupt the organizer region of the host—and a tadpole with two different heads will develop.

The experiment became Mangold's PhD thesis at the University of Freiburg in Germany. Her advisor, Hans Spemann, gave her a grade of 1–2 (not the highest grade available, but very near it). "The large, especially technical, difficulties were overcome by Mrs. Hilde Mangold with rare dexterity and perseverance," wrote Spemann. "The positive result of the experiment is of great theoretical importance." Spemann had directed Mangold to the experiment and he was her advisor, so he slapped his name next to hers on her paper.

Mangold's thesis was published in 1924. It would go on to earn Spemann the Nobel Prize in Physiology or Medicine in 1935. (Not bad for a not-quite-perfect grade.) Mangold's name was mentioned in Spemann's speech, but she was not present. She died in 1924 when a gas heater exploded in her kitchen.

Mangold's name nearly disappeared in the fire. It took sixty years, but an old friend and former colleague finally swooped in. "Very few of her contemporaries are still alive," wrote Viktor Hamburger in 1984. "As one of them who knew her well, I feel I should rescue her from oblivion."

Mangold and Hamburger both began studying at the University of Freiburg in 1920. Both came from small towns and fairly well-off families. They enjoyed hunting for orchids and shared a diverse taste in literature. Mangold could plow through books on philosophy, art history, and chromosomes in rapid succession, and she liked to share with Hamburger what she'd been reading. On the second floor of the Zoological Institute, their lab tables were next to each other. When their cytology class required donated specimens, Mangold and Hamburger traipsed around a field plucking up live grasshoppers for dissection. They also spent long hours working with embryos in the lab. "We cared more about food for thought than about nourishment for our bodies," wrote Hamburger. Food supply issues might have played a part in his assessment. Hamburger admitted that during one winter, the students survived on almost nothing but turnips.

Because it was Spemann's lab, he assigned the theses. Most of the students were given a timely project that would support Spemann's own work. When he asked Mangold to reprise an experiment from the eighteenth century, she felt she had been assigned a project only tangentially relevant. Her initial goal was to flip a hydra—a tiny, transparent, treelike polyp—inside out. Spemann wanted to know: would the hydra's insides, when flipped, act like its outsides and vice versa, as the previous study suggested?

Mangold tried and tried, without any success. Spemann even stepped in to give it a go himself but also failed. Frustrated with her lack of progress, Mangold pressed Spemann to give her another problem to frame her thesis around. As it turned out, switching thesis subjects was like coming off the bench to become a starter. It was during the new project that Mangold found the elusive organizer.

She was pleased with the experiment, but miffed when Spemann added his name to her paper as the first author—especially since Hamburger and the others were allowed to publish without such intrusion. Spemann did direct her to the organizer experiment, but Mangold was right to be miffed about the addition to her byline. Her thesis may have "initiated a new epoch in developmental biology," but over the years it's been awfully hard to keep her name attached to the accomplishment.

CHARLOTTE AUERBACH

1899–1994

GENETICS • GERMAN

TO MAKE SURE THE FRUIT FLIES RECEIVED SUFFICIENT EXPOsure, Charlotte Auerbach lowered vials of bugs into an open container where liquid mustard gas was heating. She, her research partner, and a few assistants, all of them from the University of Edinburgh, took turns exposing the specimens. "We used up a lot of technicians," Auerbach remembered. "All of them got allergic to mustard gas." After a while, who handled the flies had more to do with who was left than whose turn it was to do it. Hands burned by the task, Auerbach was eventually warned that if she didn't quit exposing her limbs to the fumes, she would cause herself serious injury.

It wasn't as if Auerbach and company weren't aware that there might be consequences. The reason they started the research in the first place was that the United Kingdom's War Office wanted to understand how mustard gas affected the body. Doctors saw patients with mustard-gas-related injuries that persisted years and even decades after they were first exposed. The goal in 1940, when Auerbach and her research partner J. M. Robson started the study, was to see if mustard gas caused genetic mutations.

From their makeshift mustard-gas lab set up on the roof of the university's pharmacology department, Auerbach carted the flies back over to the Institute of Animal Genetics to conduct a series of tests. There she examined the male flies' X chromosomes for genetic mutations. Within two months of lab work, Auerbach had a steady stream of consistent data. Confident in her results, in June 1941 she wrote a letter to the project's advisor (and future Nobel Prize winner) Hermann J.

Muller to share the news. Although the words "mustard gas" were omitted (the research was classified), Muller immediately understood the message's meaning. On June 21, 1941, she heard back. "We are thrilled by your major discovery opening great theoretical and practical field [*sic*]. Congratulations you and Robson." Auerbach later told a biographer that Muller's reply was "her greatest reward."

It was a good thing, too, as her "major discovery" wouldn't receive any other accolades for years. The results were classified and therefore unpublishable until after the war. She dangled a few clues in the journal *Nature*, but it wasn't until 1947 that she and Robson were able to publish the full account of their results. When other scientists came forward with tales of mutations caused by formaldehyde and urethane, Auerbach found herself out front in a new area of study. She was thrilled when, upon meeting a Russian visitor for the first time, he announced, "You are the mother of mutagenesis, and Rapoport [who studied formaldehyde] is the father!"

However, the attention Auerbach attracted from the scientific community for her discovery permanently ended the relationship with Robson, her partner in the mustard-gas research. Even though it was she who handled all the genetics work, he felt that he wasn't getting enough credit, particularly when Auerbach received the Keith Prize in 1948 from the Royal Society of Edinburgh. Robson believed Auerbach should have declined the award or insisted that it be shared. Auerbach pleaded with Robson to forgive her, explaining that she accepted the honor primarily because she made so little money at the university that she was in desperate need of the fifty-pound award.

Auerbach's road to recognition had been a rough one. She was woefully underpaid, and a German immigrant in Scotland who was at risk of losing her status at the university because of the war. She had just narrowly escaped Hitler's grasp.

Two experiences brought her to biology. The first, when she was fourteen, was a single, off-book, hour-long lesson by her teacher on the basics of chromosomes and mitosis. Auerbach remembered it as "one of the few great spiritual experiences of my school life." Then, while attending the University of Berlin as an undergraduate, Auerbach attended a pair of biology lectures that brought back that same sublime feeling.

Auerbach seesawed between furthering her own education and teaching secondary school to improve her financial situation. In 1933 the paper announced that she would lose her teaching position because she was Jewish. Sensing that the worst was yet to come, Auerbach's mother urged her to leave Germany. A friend of her father's had a connection at the University of Edinburgh, so Auerbach fled to Scotland. Getting into the country and settling at the university proved difficult. She was broke and, because of a paperwork snafu, was nearly rejected as a PhD student. Even after gaining her degree, she was told she had to get a job elsewhere. Auerbach appealed to F. A. E. Crew, the head of the Institute of Animal Genetics, to let her stay on. Reluctantly, he picked her up as his "personal assistant."

Crew liked his underlings to be around at all hours, but in return he offered a workspace with Ping-Pong, coffee, and an active lab-based social scene. As Crew's assistant, Auerbach did research and wrote papers. She earned a shared byline, but the position paid her only a pittance. To scrape together enough money to afford her spot in immigrant housing, Auerbach took whatever supplemental scraps of employment she could get, including tidying rodent cages, teaching, translating, and assisting in other departments.

When Crew initially assigned her to Muller, Auerbach balked. "No, I'm sorry, I'm no good at cytology." "Well, you are my private assistant," Crew replied. "And you will have to do what I tell you." Though her boss's argument wasn't very convincing, Muller (after he poked his head back in her office to apologize for the scene) offered a more compelling case. If she wasn't game, that was fine, but here's what she would be missing: though her main interest was in developmental problems tied to genes, to understand a gene, he explained, is to understand its mutation. Auerbach was in.

Just a few mustard-gas burns and some lab work later, and Auerbach was at the top of the field, the so-called mother of mutagenesis.

BARBARA MCCLINTOCK

1902–1992

GENETICS • AMERICAN

AT THE UNIVERSITY OF MISSOURI, BARBARA MCCLINTOCK, AN acclaimed geneticist working on how one generation of corn passes its genetic traits on to the next, was known as a troublemaker. The marks against her—wearing pants in the field instead of knickers, allowing students to stay in the lab past their curfew, managing with a firm, no-nonsense style—were practical choices, ones McClintock believed would improve her work and that of others. But to her superiors, her behavior was obstinate. McClintock was excluded from faculty meetings, her requests for research support were denied, and her chances for advancement were made clear: If she ever decided to marry, she'd be fired. If her research partner left the university, she'd be fired. The dean was just waiting for an excuse.

There are times for perseverance and there are times to get out quick. In 1941, after five years at the University of Missouri, McClintock found the door, slamming it behind her.

Never one to be burdened with possessions (or weighed down by the limited vision of others), McClintock hopped in her Model A Ford and, like a dandelion seed surfing the breeze, set out not knowing where she and her masterful canon of genetics work would land. When she turned her back on the University of Missouri, it was possible she was also losing the career that she'd worked so hard to cultivate.

But freedom felt like home to McClintock. When she was a baby, her mother used to set her on a pillow and leave her to amuse herself. Simply mulling over the world and all of its amazing patterns and peculiarities was a happy pastime of Mc-

Clintock's earliest years. "I didn't belong to that family, but I'm glad I was in it," she said. "I was an odd member."

Her outsider status was not so different in the scientific community. Though she absolutely belonged there and was fully absorbed in her work, McClintock never completely integrated. One part of the issue was societal. Getting a faculty position at a university was exponentially harder for women in the 1920s than it was during World War II, when positions opened up for women when men were called to war. Though up to 40 percent of graduate students in the 1920s in the United States were women, that didn't translate into jobs—especially in science. Fewer than 5 percent of female scientists in America were able to land jobs at coed institutions. And even then, the home economics and physical education departments were the biggest hirers. Women rarely rose to posts as prestigious as professor. In the Venn diagram of female biologists hired as professors at major research institutions, the middle was a lonely place. McClintock never got there.

McClintock's work also kept her out of the mainstream. She was either ahead of her time, with experimental methods so dense and complicated that they were difficult for her peers to understand, or she chose subjects that operated outside trends in biology.

During her first year of graduate school at Cornell University, for example, McClintock took it upon herself to identify discrete parts of corn's chromosomes. Her short-term advisor, a cryptologist, had been after the same tricky-to-find prize for a long time. McClintock saddled up to the microscope and—bam—"I had it done within two or three days—the whole thing done, clear, sharp, nice." She revealed the answer so quickly that it bruised her advisor's ego. McClintock was so thoroughly hopped up on the quest that she hadn't even considered the possibility that she would upstage her superior. In other instances, her groundbreaking experiments required an interpreter. When she laid out her case for the location of genes on corn's distinguishable ten chromosomes, her method remained a mystery to her colleagues until a scientist from another school visited and unpacked the study design for public consumption. "Hell," said the interpreter. "It was so damn obvious. She was something special."

McClintock adored biology at Cornell. She was no typical high achiever. Following the acknowledgment of her corn chromosome discovery as a master's student, she attracted a pack of professors and PhDs who trailed her around campus, "lapping up the stimulation she provided," said one, like puppies tumbling after castoff treats. Together the group, with McClintock as its intellectual leader, ushered in an especially bright period of genetics. McClintock proudly recounted how the "very powerful work with chromosomes . . . began to put cytogenetics, working with chromosomes, on the map. . . . The older people couldn't join; they just didn't understand. The young people were the ones who really got the subject going."

Post-PhD, McClintock spent a few more years at Cornell, publishing papers, teaching botany, and advising students. In 1929, she and a graduate student bred together one strain of corn with waxy, purple kernels with another strain that had kernels that were neither waxy nor eggplant-colored. McClintock's experiments showed that some kernels inherited one trait but not the other, for example, brightly colored kernels without the waxy texture. When McClintock looked at the chromosomes through a microscope, she found that their appearance was noticeably different, and in the cases where kernels had one trait but not the other, parts of a chromosome had traded places.

The discovery was hailed as one of the greatest experiments of modern biology. At just twenty-nine years old, McClintock proved herself a powerful force in genetics research—but without a permanent faculty position. The head of the department was in favor of bringing her on to become a professor but the Cornell faculty forbade it. So McClintock left, picking up fellowships here and there, searching for a new place to put down roots.

The country's greatest research institutions should have fought over McClintock, but instead she ended up searching for a space to plant her corn. She found one at Cold Spring Harbor in Long Island, New York. The facility was initially founded in 1890 as a place for high school and college teachers to learn about marine biology. When McClintock arrived, it was a genetics institute. The atmosphere was ideal; McClintock wouldn't have to teach, and there were no restrictions on her research, which would be entirely self-directed. She could wear jeans and stay as late and as often as she wanted. The place suited her so well that when she socialized, she would invite friends to the lab instead of to her "home," an unheated, converted garage down the street used for nothing more than sleep.

McClintock was extraordinarily organized. Clothes in her closet all faced the same direction, and each of her scientific specimens was assiduously labeled. Sometimes she'd get so engrossed in her work that peering into a microscope would feel to her like spelunking through the deep secrets of a cell. "You're not conscious of anything else," she remembered. "You're so absorbed that even small things get big."

At Cold Spring Harbor, McClintock spent six years on her greatest scientific accomplishment. When she finally unveiled her findings to a group of researchers, her hour-long talk was met with silence. One listener recalled that the talk landed "like a lead balloon." McClintock had just laid out a meticulously researched case that genetics was much, much more fluid than what scientists had previously realized, with genes able to switch on and off and change locations. The prevailing belief was that genes were like bolted-down pieces of furniture. In the 1950s, scientists from all different fields of study were getting into the genetics game; chemists and physicists applied their disciplines to understanding inherited traits. With so many new ways to look at our genetic makeup, corn had fallen out of favor. "I was startled when I found they didn't understand it, didn't take it seriously," she said of the talk. "But it didn't bother me. I knew I was right."

That she was. The acceptance of her ideas didn't come until nearly two decades later, when molecular biologists finally saw in bacteria what McClintock had seen in corn. At the news, McClintock was overjoyed. "All the surprises . . . revealed recently give so much fun," she wrote to a friend. "I am thoroughly enjoying the stimulus they provide." Public acknowledgment brought a string of awards—the MacArthur Foundation Fellowship, the Albert Lasker Basic Medical Research Award—but no Nobel. Then finally, in 1983, thirty-two years after her big-but-ignored discovery, she heard her name announced on the radio. She had finally won science's most prestigious prize. Her "discovery of mobile genetic elements" was touted by the Nobel Committee as "one of the two great discoveries of our times in genetics."

In the ensuing years, she was asked time and time again the same question, some delicately worded take on *Were you bitter it took so long?* Her answer: "No, no, no. You're having a good time. You don't need public recognition, and I mean this quite seriously, you don't need it." With characteristic confidence, she added, "When you know you're right you don't care. It's such a pleasure to carry out an experiment when you think of something. . . . I've had such a good time, I can't imagine having a better one. . . . I've had a very, very satisfying and interesting life."

SALOME GLUECKSOHN WAELSCH

1907–2007

DEVELOPMENTAL GENETICS • GERMAN

SALOME WAELSCH SAW "FURTHER AND WIDER THAN ALMOST everybody." At lectures, she was known for contextualizing a new scientific finding by placing it within the history of biology or genetics on the fly. Some of the most brilliant scientists of her generation had a hard time seeing outside the bounds of their specialty. Waelsch's vision was expansive. By stitching together two fields—genetics and embryology—she created a new one. In 1938, Waelsch co-founded developmental genetics, a field dedicated to the role of mutation in gene development, and explained its methodology.

Her place as a founder was deeply tied to her personal philosophy. "I am convinced that scientists do not operate intellectually or experimentally in a vacuum totally divorced from personal, social, and political phenomena in their environment," Waelsch explained. "I have always paid attention to these extraneous factors."

Before Waelsch was born, her parents moved from Russia to Germany. Being Jewish, they faced significant prejudice. They instilled in their children the importance of education, which, Waelsch pointed out, "help[ed] me more than I could have imagined, because I later became a Hitler victim." Education gave Waelsch a ticket out of Germany when her life depended on it.

After initially considering a degree in Greek and Latin—and being talked out of it by friends who questioned a language degree's utility—Waelsch, who had always found science utterly enthralling, decided to switch to chemistry and biology.

In 1928, Waelsch moved from the University of Berlin to the University of Freiburg to get her PhD. When she arrived, she reckoned her advisor (and soon-to-be Nobel Prize winner) Hans Spemann just didn't have the courage to tell her she couldn't study there. "Our first meeting made it quite clear we were not meant for each other," recalled Waelsch. Friends of hers thought she might have been too outspoken for his tastes; Waelsch felt that he didn't respect women equally. In any case, Spemann's lab was a hugely important place for experimental embryology. Though she admired his work, Waelsch wished she could have played a larger part in it. While others were given the meatiest problems in the field, Waelsch was given the supporting work, just the scraps. Despite the rift, Waelsch learned an extraordinary amount from Spemann. As was her way, she absorbed everything she could from working with him, knowing it would help her later.

Being the global expert on experimental embryology, Spemann was a major draw for visiting researchers. Waelsch was savvy enough to form meaningful relationships with those who passed through. During her three years at Freiburg, she hit it off with Viktor Hamburger, who both supervised her dissertation research and introduced her to genetics, as well as a fellow student named Conrad Waddington, who showed her how genetic and developmental research were related.

As she was close to the point of establishing her own career, Waelsch's husband was fired from his university position because he was Jewish. Having already made a name for himself as a promising biochemist, he was immediately offered a job at Columbia University in New York. He and Waelsch moved from Germany to the United States shortly thereafter.

In 1936, after three years without a job, Waelsch met L. C. Dunn, a geneticist at Columbia, at a faculty social gathering. He was looking to hire a developmental biologist, so they made a deal to bring her on. It was a deal, not a hire, because there were a few caveats including that he couldn't compensate her with a salary. But for her borrowed expertise, he would trade her training in genetics research. Waelsch desperately wanted to get back to work, and that proposed combination of specialties struck her as particularly appealing. She agreed to join him.

Although the pay was nonexistent, the environment suited her. Within two years, Waelsch published one of her most important works. In the introduction (the paper was based around research on tailless mice), she outlined not only the goals of the new field of developmental genetics but also its research methodology, which was markedly different from anything else before it. In experimental embryology, for instance, scientists designed an experiment to test a hypothesis. In developmental genetics, Waelsch explained, a naturally occurring genetic mutation "carries out an 'experiment' in the embryo by interfering with the normal development." In mutation-led studies, the scientist looked at the whole chain of development, from an abnormality in the DNA all the way to its physical conclusion. Waelsch wrote, "The developmental geneticist first has to study the course of the development (that is, the results of the developmental disturbance) and can then sometimes draw conclusions on the nature of the 'experiment' carried out by the gene." This declaration served as a call to action for a brand-new field.

Waelsch's tenure at Columbia would be a very productive nineteen years, though she didn't have a faculty appointment. Whenever she asked to become a larger part of the Department of Zoology, she was given all sorts of creative excuses that could be chalked up to the same thing: not a woman, not now.

In 1955, she was invited to become a faculty member at a new university, the Albert Einstein College of Medicine, in New York. Genetics didn't really have a place in university education at the time, so Waelsch was called upon to teach some of the school's earliest genetics classes. Although she started as an associate professor, Waelsch rightfully zipped up the ranks, eventually becoming chair of genetics.

Combining her deep knowledge of science history with her experience in the field, in 1992 Waelsch gave a talk covering the previous fifty years of research in developmental genetics. She led the audience back to her formative years, explaining how Hamburger and Dunn had helped the new field take shape. She guided the audience through the major milestones—nerve growth factors and regulatory genes—to the present day. As she neared the end of the talk, she left the future of the field open. Just as Waelsch had combined disciplines to make her own, she looked to the researchers of the future to come up with new, interesting combinations of specialties. "I personally am increasingly impressed with the degree to which molecular developmental biology and molecular genetics are merging into one science." In other words, the future is malleable, and to see it, you just have to listen to history and have a grand enough vision.

RITA LEVI-MONTALCINI

1909–2012

NEUROEMBRYOLOGY • ITALIAN

DURING THE LAST TWO AND A HALF DECADES OF HER 103 YEARS, Italians liked to joke that everyone would recognize the pope, so long as he appeared with Rita Levi-Montalcini. Though she stood only five feet, three inches, the stories of her work and her life were as large and dramatic as her iconic sideswept hair.

There was the time she smuggled a pair of mice on a plane to Brazil by tucking them away in her purse or pocket—for the sake of her research, of course. Or the years she bicycled door-to-door during World War II, pleading with farmers for donated chicken eggs to feed her "babies" (they were actually embryonic research). The plight was a ruse. She needed fertilized eggs for her work. Once, Levi-Montalcini talked her way into the copilot seat on a fully booked flight. On another flight, when the airline lost her suitcase and the clothes she had on were wrinkled, she opted to give a lecture in a pressed nightgown rather than appear disheveled in front of an audience.

In life and in her work, Levi-Montalcini preferred grand gestures and big risks. As a child, she vowed never to marry, in order to devote herself completely to science—a promise she kept. Finishing school? No thanks. She was meant for medical school. When the Italian government barred her from medicine and research in 1938 because she was Jewish, she set up a secret lab in a bedroom so she could continue to examine the development of fibrous nerve cells, an interest she cultivated while working toward her medical degree.

During this time, Levi-Montalcini read an article written by the founder of developmental neurobiology, a German embryologist based in St. Louis, Missouri, named Viktor Hamburger.

Hamburger used chick embryos to inquire about a possible link between the spinal cord and the development of the nervous system. The idea piqued her interest. Even while operating undercover, Levi-Montalcini figured she could talk her way into a regular supply of chicken eggs.

Levi-Montalcini sprang into action, conducting her own experiments to see if she could suss out a link. She recruited a similarly ousted professor as a research partner and called on her family to provide lab support. Her brother built an incubator for the eggs she gathered, and Levi-Montalcini made a scalpel from a filed-down knitting needle. She also acquired a slew of tiny instruments, like forceps made for a watchmaker and scissors made for an ophthalmologist. She used these miniature tools to extract the chick embryos and cut their spines into thin slices. After studying the neurons in the spinal cord at different stages of embryonic development, Levi-Montalcini discovered something entirely new. Nerve cells didn't fail to multiply as previously thought; they both grew and died as a normal part of the development process.

Because she couldn't publish in Italy, Levi-Montalcini sent her papers to Swiss and Belgian journals available in America, which is where Hamburger learned about her work. After World War II had concluded and Levi-Montalcini was allowed to conduct scientific experiments outside the bedroom, Hamburger invited the Italian researcher to come to Washington University in St. Louis to discuss their overlapping interests. She accepted, and a trip that should have taken a few months turned into a twenty-six-year tenure at the institution.

With Levi-Montalcini's knowledge of the nervous system and Hamburger's foundation in analytical embryology, the pair was ideally matched to tackle the mystery of how nerve cells emerge and extinguish together. Levi-Montalcini thrived in her new environment, working extraordinarily hard from morning until late into the evening. Despite the work experience, Levi-Montalcini still believed that her biggest accomplishments were guided by intuition. "I have no particular intelligence," she said. But when the powerful weathervane inside her landed on a direction or a thought, "I know it's true. It is a particular gift, in the subconscious. It's not rational." Hamburger leaned toward crediting talent. "She has a fantastic eye for those things in microscope sections . . . and she's an extremely ingenious woman."

Levi-Montalcini traveled to Brazil to learn how to grow tissues in a glass dish. On this trip, however, the scientist was failing. Trying the method she'd learned, Levi-Montalcini swung back and forth between enthusiasm and despair. (Even her mood swings were legendary.) By relying on mice to produce nerve cells, researchers were locked into a set timeline of growth. But if Levi-Montalcini could make those special cells in the lab, her experiments would accelerate. Still . . . the technique wasn't working. On her final attempt, Levi-Montalcini plopped a scrap of embryonic chick cells on one side of a petri dish and a chunk of tumor on the other. When placed next to each other—but not touching—the nerve fibers astonishingly started to stretch, extending out from the cells in every direction like a fragile, otherworldly crown. It was an extraordinary show indeed—and one Levi-Montalcini took pleasure in playing over and over again throughout her career.

What was the factor shoving nerve growth into action? Upon her return to St. Louis, Levi-Montalcini figured it would take her a few months to find out.

A few months went by . . . and then a year, two, three; all the while she and her then research partner Stanley Cohen working furiously. (By that point, Hamburger had stepped back from research and become more of a mentor.) The team grew tumors, experimented with snake venom, and spent a lot of time thinking about mouse saliva. It took six years, until 1959, to identify the nerve growth factor in a mouse's salivary gland and purify it into something that would trigger that ethereal crown.

At one time, the discovery was seen as a small thing, impressive but also niche. But as more and more growth factors were discovered, the field bloomed. It was found that nerve growth factors influence everything from degenerative disease progression to the success of a skin graft to damaged spinal cord protection.

In 1986, she and Cohen were awarded the Nobel Prize in Physiology for their work.

The prize launched Levi-Montalcini into Italian celebrity. (She had returned to Italy part-time in 1961.) In her later years, she took work calls on the car phone as a driver chauffeured her around in a Lotus. Levi-Montalcini was awarded the National Medal of Science, and in Italy she was appointed a senator-for-life. "The moment you stop working," she said, "you're dead." Wearing a string of pearls, high heels, and a broach under her lab coat well into old age, she made it to 103.

ROSALIND FRANKLIN

1920–1958

GENETICS • BRITISH

THE DISCUSSION OF ROSALIND FRANKLIN'S LIFE AND WORK often rotates around one impossible question: Had she not died of ovarian cancer at the age of thirty-seven, would she have shared the 1962 Nobel with James Watson and Francis Crick? That answer is probably not.

The conclusion stings because there was some definite wrongdoing. In Watson's bestselling book *The Double Helix*, which recounts his and Crick's discovery of DNA, Watson caricatured Franklin cruelly. She was "Rosy" (a name she did not like), who "might have been quite stunning had she taken even a mild interest in clothes." *Rosy*, who was curt and reactive and caused everyone working with her misery. *Rosy*, who could not possibly be considered serious competition in the quest to nail down the structure of DNA.

Because she had been dead for a decade when *The Double Helix* was published, others spoke for her. It was "a mean, mean book," remembered the Nobel Prize–winning geneticist Barbara McClintock. Another geneticist, Robert L. Sinsheimer, called Watson's portrait of Franklin "unbelievably mean in spirit, filled with the distorted and cruel perceptions of childish insecurity." Anne Sayre, a friend and Franklin biographer, complained that Watson had "carelessly robbed Rosalind of her personality."

Watson's portrayal of Franklin, however, was made worse by this cavalier disclosure: Rosy "did not directly give us her data." And there it was, a stunning admission hidden between chapters of gloat. When others tugged on the dangling thread, the portrayal of Franklin began to unravel. Watson may have found her someone unpleasant to work with, but his experience was by no means universal. She was a competitor—and far ahead of Watson and Crick during much of the search for DNA. The rival pair simply wouldn't have made their discovery when they did had it not been for two crucial pieces of information passed from Franklin's lab at King's College in London to Watson and Crick's at Cambridge without her knowing it.

The first: a clear photo of the structure of DNA, calibrated and captured by Franklin. The second: an internally circulated report that recapped the results of her recent work. Watson and Crick had already made some headway into the structure of DNA, but they had gotten the water content and the location of the phosphate sugars wrong. Without Franklin's data, they wouldn't have had the essential pieces they needed to solve the puzzle. Franklin eventually would have come to thc same conclusion as Watson and Crick—the helix, the base pairs, the direction of the phosphate chains—some say, had her work not been shared.

"All her life, Rosalind knew exactly where she was going," her mother recalled. Once her mind latched onto something, she was all in. At age six, Franklin was described by her aunt as "alarmingly clever. . . . She spends all her time doing arithmetic for pleasure, [and] invariably gets her sums right." Franklin was precise, literal, and always more at home with data than with speculation.

While Franklin was studying at Cambridge, her father complained that she felt about science as she should about religion. Franklin held her ground. "You frequently state . . . that I have developed a completely one-sided outlook and look at everything and think of everything in terms of science," she replied in a letter. "Obviously my method of thought and reasoning is influenced by a scientific training—if that were not so my scientific training will have been a waste and a failure. . . . Science and everyday life cannot and should not be separated."

How could she contribute to the World War II effort, since her father insisted? Science was the no-brainer. Following her graduation from Cambridge in 1941 and a research position, Franklin bicycled daily across prime air raid territory to a post she'd found at the British Coal Utilization Research Association. There her job was to figure out why some kinds of coal allowed gas and water to filter through and why others put up a more efficient blockade. (Charcoal had been used in gas masks, so it was important wartime research.) Franklin published five papers on the material's properties by the time she was twenty-six. Her thesis, which covered "solid organic colloids with special reference to coal and related materials," earned her a PhD. Additionally, her research in the 1940s would help advance the development of carbon fiber later on.

After the war, a friend recommended her for a job in Paris as a physical chemist, again working on coal. The three years she spent abroad were perhaps her happiest. She made friends, spoke the language flawlessly, and felt more at ease in her surroundings than she ever had at home. Tugged back to England by the feeling that London would accelerate her career, at age thirty Franklin returned to the UK.

She began work at King's College in London upon her arrival. There she took over the study of DNA, originally initiated by an interdisciplinary team that had set it aside for the better part of a year. The goal was to figure out DNA's molecular structure. To do so, Franklin lined up DNA fibers, bundled them together, and X-rayed the carefully prepared samples in 75 percent humidity and 95 percent humidity. At 95 percent, the molecules elongated, which Franklin called DNA's B-form. The pictures of DNA in this state looked like the lines of an X blinking in and out of focus—the sign of DNA's helical structure, though she didn't yet know it.

At King's College, Franklin didn't have any formal collaborators. The most obvious choice would have been Maurice Wilkins, also at King's, but an early misunderstanding about Franklin's role turned the colleagues into adversaries. Their relationship had consequences for Franklin when Wilkins, complaining to Watson about his colleague, pulled out her beautiful B-form and shared it with the American working at Cambridge without her approval.

This photograph—taken by Franklin—was a major revelation for Watson, who had been working from muddy images that were a mix of DNA's dry and wet forms. Franklin's clear image of DNA's wet form changed the way Watson and Crick understood DNA.

Watson and Crick's next breakthrough also came thanks to Franklin, and again without her knowledge. In 1952, Franklin was asked to summarize her previous year's work for a government committee. Max Perutz gave her summary to Watson and Crick. (The paper was not marked confidential, but the report also wasn't intended for any eyes outside the committee.) The report gave the pair from Cambridge crucial information about the dry and wet forms of DNA. Combined with their own research, Franklin's pieces were enough for Watson and Crick to form a solid understanding of DNA's structure. Announcing their discovery in *Nature*—that DNA was a helical ladder, with one side going up and the other going down—they claimed the prize for finding the solution without revealing Franklin's part in their discovery.

Franklin got scooped by the Cambridge team at the same time as she was on her way out at King's College. She felt that the environment wasn't good for her, and many of her colleagues agreed. As the discoverers were crowned, Franklin transferred to Birkbeck College and away from DNA research, as was stipulated in her transfer agreement.

At Birkbeck, Franklin set up a research group that looked

at ribonucleic acid's role in virus reproduction. For scientists studying a virus's molecular structure with X-rays, her group was the best in the world, revealing, among other things, how proteins and nucleic acids fit together to transmit genetic information. To study polio, Franklin convinced the wife of a colleague to sneak the virus in a thermos from the United States to London on a plane.

Despite problems with Watson, Franklin became good friends with Crick and his wife, who was French. In Franklin's last year alive, her work got a moment of public recognition. For the 1958 Brussels World's Fair, she constructed a massive six-foot-tall display of the tobacco mosaic virus, a pathogen that affects hundreds of different plants.

Word of Franklin's essential part in the discovery of DNA did not get out until Watson himself spilled it. Since then, she's become the subject of several biographies and a poster child for those who didn't receive the credit they deserved. Franklin, who was always deeply invested in data and facts, would have been happy to know that so many people cared about her concrete contributions.

ANNE MCLAREN

1927–2007

GENETICS • BRITISH

ANNE MCLAREN WAS NOTORIOUSLY CAGEY ABOUT HER PAST, not because it was painful, but because it wasn't: her family was wealthy, and she enjoyed a fine upbringing. Her scientific contributions were impressive—she was an in vitro fertilization trailblazer and the first person to bring a test tube baby mouse into the world—and if there had to be attention, she'd rather it be on her work than on herself.

The story of McLaren's career can be effectively told through her mice. In 1955, she had too many of them. At the time, McLaren had taken a fellowship at University College London. Fresh from a PhD in zoology from Oxford three years prior, McLaren bred mice to observe how uterine environmental factors influenced embryonic development. As she quickly discovered, the environment wielded a lot of influence.

McLaren and her research partner (and then husband) Donald Michie churned out a series of experiments with intriguing results. In one, they found that a strain of mice typically born with six lumbar vertebrae would be born with five if the six-lumbar embryo was implanted into a host whose genetic strain had a lower number. The environment of a different strain's womb nudged the foreign embryo to take on some of its characteristics.

To carry out her research in embryonic development, McLaren needed her mice to produce a generous supply of eggs. Waiting for natural biological systems like gestation is time-consuming—about twenty days for mice—so while working on other projects, McLaren figured out how to give the existing procedures promoting super-ovulation (how scientists coerced mice into releasing more eggs at once) a substantial boost. When she needed a faster, more efficient way to transfer embryos between mice, she developed one. These were extremely productive years for both McLaren and her mice. The space at University College simply couldn't keep pace with her research—or rather, with the sheer number of cages it required.

In 1955, McLaren, Michie, and their mice all moved to the Royal Veterinary College in Camden Town, London, where they worked out of what was called the "canine block." McLaren's new lab was twenty-five feet by twenty-five feet, with a tiny office in one corner. Her rodent research subjects had their own digs on an upper floor.

Fascinated by "everything involved with getting from one generation to the next," McLaren began looking into how inbreeding influenced a mouse's morphology, and then she tested how extreme environmental temperatures impacted embryonic development. The latter experiment was carried out in carefully temperature-controlled rooms on the roof of the Department of Hygiene. Gestating mice were placed either in hot rooms, in rooms with average heat levels, or in cold rooms. The litters in average and hot temperatures produced pups of normal size and weight, but McLaren found that the cold affected litters negatively. The babies were slower to grow.

McLaren enjoyed spending time with her collaborators, both mouse and human. She spent hours upstairs with her animals, recording and tagging the newborns. When downstairs in the lab, McLaren even used a typewriter with an exercise wheel affixed to the top of it. As she put her research in writing, her mice would jog in the wheel above.

In the mornings, McLaren took her coffee into the Senior Common Room of the Veterinary College. There she would meet up with other researchers in the building and chat about the latest developments in their fields. One morning in 1956,

McLaren, Michie, and John Biggers, a cell biologist studying the use of embryonic chicken bones for organ cultures, started discussing a new paper in *Nature* that reported that eight mice embryos had been cultured to the blastocyst stage, an early stage of embryonic development. As they spoke, the three scientists realized that together they had all the expertise and background necessary to take the *Nature* idea one step further to its logical conclusion: the successful birth of mice born via in vitro fertilization.

They jumped into the project immediately. Biggers repeated the process that was published, culturing mice embryos in test tubes to the blastocyst stage before handing the work over to McLaren, who transferred the blastocysts into surrogates. Then they waited. Biggers was away on vacation when McLaren got the results, which she reported via telegram. The news of their success delighted her research partner but thoroughly weirded out the post office staff. It read: "Four bottled babies born!"

The team's breakthrough experiment would ripple outward for decades. Between 1982 and 1984, McLaren was asked to be a part of the Warnock Committee, a group tasked with drafting the first set of ground rules regarding in vitro fertilization of human embryos. McLaren was the only participant with specific experience, and her guidance and explanations of the science—delivered with "impeccable clarity"—gave the report legs. The guidelines she helped develop became the standard and a model for other countries to emulate. She not only proved in vitro fertilization was possible, but years later, she was also responsible for safely and ethically guiding it into the world.

McLaren's work was further celebrated when she was given the female equivalent of a knighthood. In 1993, she was honored with the title "Dame."

LYNN MARGULIS

1938–2011

BIOLOGY • AMERICAN

IN THE EARLY 1990S, BIOLOGIST LYNN MARGULIS WAS AT A DINner party where a number of similarly accomplished scientists were in attendance. Margulis had made an early name for herself with an unpopular theory about the origin of eukaryotic cellular life. The dinner came directly on the heels of harsh criticism of her work published by another prominent biologist—a biologist who happened to be there. Margulis addressed her critic directly at the dinner, picking apart each point of his argument and countering it with a spirited defense. Soon she had the biologist cornered. The theoretical physicist Lee Smolin, an admirer of Margulis's work, later compared her rebuttal to Aristotle defending the Copernican theory at dinner parties in Rome: "I saw in her the same confidence in her vision, together with impatience at those who can't think as openly or as broadly but instead choose to misunderstand the new ideas."

Margulis always prided herself on her ability to see through perceived wisdom—even as a child. She regarded her teachers' orders and their justification—*because I said so!*—with skepticism. When faced with dull schoolwork, she often chose punishment over participation. What took her from a skeptical child to a discerning scientist was learning about the experimental method and being sent outside to observe the world.

As Margulis recalled in an interview at Rutgers University in 2004, "Science was a way to find out directly about the world from evidence. I had never seen that in my life." Margulis realized that she didn't need textbooks or teachers to filter information for her. She could find answers from the world itself. And she did. Margulis watched ants march through blades of grass to learn about their behavior. "It just felt right to go into nature. Always." In high school at the University of Chicago's laboratory school, Margulis was encouraged to read the works of Isaac Newton and Gregor Mendel instead of textbook summaries. She later recalled, "Classes were not required. And that's why I went to them all."

Margulis earned an undergraduate degree from the University of Chicago at eighteen and a PhD in genetics from the University of California, Berkeley, a few years later. By 1966 she'd landed a position as a biology instructor at Boston University. It was there that her boldest ideas took shape. Margulis had long been fascinated with the mitochondria in eukaryotic cells, the kind of cells that contain a nucleus. She thought that the sausage-shaped organelles, which serve as a cell's power plant, looked a whole lot like bacteria. She wasn't the first to notice the similarity; others had written about the resemblance before and been ridiculed for it. Undeterred by their failure, Margulis decided to flesh out a more compelling theory.

Long, long ago, Margulis postulated that one bacterium would have to have housed another independent bacterium. But instead of one extinguishing the other, Margulis believed something beautiful happened: the two struck a deal. Together they gained advantages, like speed and appetite. The descendants of the partnership became a plant cell's chloroplasts and an animal cell's mitochondria. The cooperative actions of bacterial cells was the reason animals started to suck in oxygen and plants were able to convert light into energy.

Margulis was just two years past her PhD—still a rookie—when she sent her paper out for publication. Her interests were radically different from the accepted evolutionary theories of the day, which focused on competition as the primary driver, not cooperation. And while almost everyone else relied on animal fossils for research fodder, Margulis felt that by looking back to our oldest cells and microorganisms, she focused on a more accurate, historical account of life on Earth.

Fifteen academic publications rejected her paper before the *Journal of Theoretical Biology* finally accepted it in 1967. Even after her work went public, Margulis's ideas were deeply unpopular in the scientific community, sometimes even drawing hostility. But instead of backing down, Margulis went all in, building her paper into a book called *The Origin of Eukaryotic Cells.* When the book was released in 1970 by Yale University Press, the American philosopher Daniel C. Dennett remembered in *The Third Culture: Beyond the Scientific Revolution*, "she was scoffed at, laughed at." Eight years later—a decade after her original paper came out—new research was published offering concrete support for Margulis's radical ideas. A former student recalled Margulis strutting into her classroom, paper in hand, with the wide smile of sweet validation. "It's delicious that this is now pretty well accepted as a major, major theoretical development," said Dennett. "I think of her as one of the heroes of twentieth-century biology."

Margulis had rewritten the book on evolution, and today it's "one of the classics of biology in the twentieth century," according to the Chilean biologist Francisco Varela.

Not all of Margulis's ideas gained acceptance, though. Later in her career, Margulis had a long-running academic partnership with the British scientist James Lovelock, whose so-called Gaia hypothesis, which proposed that Earth is one self-regulated organism, is still widely criticized. When asked about her legacy by *Discover* magazine, Margulis explained, "I don't consider my ideas controversial. I consider them right."

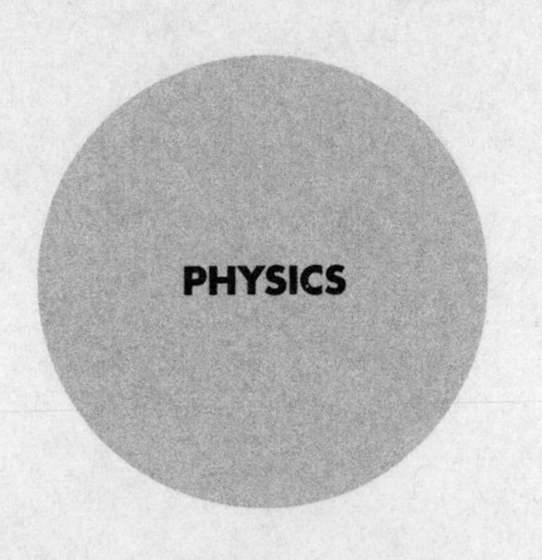
PHYSICS

ÉMILIE DU CHÂTELET

1706–1749

PHYSICS • FRENCH

FOR MANY, MANY YEARS, THE ONLY COMPLETE FRENCH TRANSlation of Sir Isaac Newton's masterpiece, *Philosophiae Naturalis Principia Mathematica* (or *Principia*), was written in 1749 by Émilie du Châtelet (Emilia Neutonia, to those poking fun). It took her four years to complete the project, which included the *Principia* translation, 287 pages of additional commentary, and a new mathematical addendum. Although she'd been working on it for years, the final months leading up to her deadline were particularly hectic, with du Châtelet putting in seventeen-hour workdays that often ended at 5 a.m. When sleep encroached, she dipped her limbs in ice water and gave her arms a good smack to refocus. Scribbling so rapidly that she only sometimes lifted the quill up between words, du Châtelet's fingers were often stained with ink. Her deadline was to be before the birth of her fourth child. She finished her beast of a manuscript and gave birth to a daughter in the same week. Du Châtelet died ten days later.

The death of du Châtelet, considered one of Europe's top savants, a "phenomenon," and "a genius worthy of Horace and Newton," at age forty-two was a great loss to the scientific community. However, in a relatively short amount of time, she contributed a great deal to her contemporaries' understanding of Newtonian physics and helped generations of scholars make their way through Newton's famously dense tome.

How du Châtelet acquired the academic foundation for her work isn't clear, as the details of her upbringing aren't well documented. Historians have only a rough outline: a father who was an official in Louis XIV's court, a marriage arranged for du Châtelet at age eighteen, an older husband of noble lineage who worked in the military, children born, houses maintained. Then, at the age of twenty-six, du Châtelet started taking private lessons in mathematics—advanced geometry and algebra—with a prominent mathematician and thinker, and then with another teacher, a young mathematics prodigy. While many French academics were loyal to René Descartes, both of du Châtelet's teachers were loyal to Newton.

For du Châtelet, learning about mathematics and Newtonian physics was like putting on a pair of prescription glasses for the first time; all at once, she could see the individual leaves on a once-blurry tree. She got a thrill from applying her new knowledge to unconventional places. She could figure out how strong a gust of wind needed to be in order to cause a branch to sway. Equations gave her the ability to map a bird's flying pattern. Along with math and physics, du Châtelet voraciously consumed texts of philosophy, literature, and other sciences. Within two years, she had leapt ahead of one tutor, and declared her home in Cirey (about 150 miles east of Paris) a welcoming place for intellectuals to visit and work. Many took her up on the offer. As du Châtelet pointed out while discussing an English poem she had translated into French, it "sometimes happens that work and study force genius to declare itself." Du Châtelet was boldly stepping out.

One of her first public acts as an intellectual was to enter an annual competition put on by France's l'Académie royale des sciences in 1737. The subject was the nature and propagation of fire; at the very last minute, just two weeks before the deadline, du Châtelet decided she'd submit a paper. What gained her entry was anonymity; the judges couldn't discriminate without a name. Though she didn't win, her work was later published with this attribution: "by a young Lady of high rank." Because there was no one else quite like du Châtelet during her time, the clues to her identity would have been obvious to her contemporaries.

Gaining confidence and momentum, du Châtelet published her first book, *Institutions de physique*, in 1740. She wanted to teach her son the fundamentals of physics, but there simply weren't any texts available that offered a concise explanation of the principles. *Institutions de physique* clearly and precisely presented the ideas of Newton, Descartes, and Gottfried Wilhelm Leibniz. Though it was a thorough synthesis, others took it as an opportunity to take her down.

Shortly after the book was released, the secretary of l'Académie royale des sciences published a letter questioning the claims of du Châtelet's text. In it, he accused women of being fickle, and du Châtelet specifically of having a feeble mind. He insisted she had misunderstood both his scientific work and her beloved Newtonian physics, and he argued that her fundamental understanding of mathematics was incomplete. The letter was brazen and patronizing. Scientific debate is one thing, but this letter was an attempt to knock her out of the conversation entirely.

Du Châtelet had far too much pride to let the baseless criticism stand without rebuttal. Her response was swift. Dismantling the secretary's letter with surgical precision, she addressed his criticisms point by point, talking down to him in the same smug tone, all the while elaborately displaying her mastery of her subject. Du Châtelet sent her detailed response to five hundred Académie members. The secretary had just recently been elected, but her letter forced him into early retirement.

Du Châtelet's greatest accomplishment—translating and annotating Newton—would be her last. The text was a natural evolution of the work she'd been doing for years. Unfortunately, du Châtelet's important contribution to physics is often overshadowed. When her name appears in books and articles, it's often a salacious aside; du Châtelet is mentioned mainly as Voltaire's mistress and muse.

Voltaire and du Châtelet lived together in her home in Cirey for some fifteen years, give or take a few stretches when they were apart—and even after Voltaire moved on romantically. (Du Châtelet's husband was not only fine with the arrangement, but he also supported the work of both his wife and her lover.) Du Châtelet's first significant scientific contribution was hidden in Voltaire's. In his book *Éléments de la philosophie de Newton,* the sections on more abstract mathematics and science were contributed by du Châtelet. By the time of her death, she had become a far more accomplished Newtonian scholar than Voltaire, and yet their relationship continues to outshine her science wherever their history is written.

Du Châtelet hurried to secure her legacy, finishing her *Principia* translation, filling out explanations for things that were missing here, and strengthening Newton's case with a new equation there. It was almost as if she knew it would be her final act.

LISE MEITNER

1878–1968

PHYSICS • AUSTRIAN

LISE MEITNER SHOULD HAVE WON THE NOBEL PRIZE. SHE WAS an exceptionally talented nuclear physicist, both systematic in her research and a deep, critical thinker. The reasons that her name did not appear with her research partner's for the discovery of nuclear fission? Well, some were political, some were situational, and others were just downright unfortunate.

In the early twentieth century, Germany was an epicenter of great scientific minds—Meitner's included. Meitner spent the majority of her career in Berlin, where she became friends with other physics all-stars, among them the Nobel Prize winners Max Planck, Albert Einstein, Niels Bohr, and James Franck. Weekly, she participated in a colloquium where around forty experts in her field gathered to present and discuss new research. The first row was reliably stacked with heavyweights, and it was Meitner's rightful place. Einstein called her "Our Madame Curie."

But then, because she was Jewish, Hitler made it the wrong time for Meitner and many of her peers to be living and working in Germany. Not wanting to abandon her projects or the scientific community in Berlin, Meitner put off leaving for years. But in the summer of 1938, with increasingly alarming restrictions placed on Jews—including barring Jewish scientists from leaving the country for conferences "where they appear to be representatives of Germany"—Meitner escaped to the Netherlands with the help of some Dutch friends who arranged for her entry into the country without a passport. Meitner left Germany, where she had lived for thirty years, with just two lightly packed suitcases. She also left behind the biggest project of her lifetime—the one she initiated in 1934 that would lead to the discovery of nuclear fission and eventually to the Nobel Prize.

Meitner had faced adversity before. In her hometown of Vienna, she wasn't allowed to go to school after the age of fourteen. But Meitner did not let this barrier impede her academic progress. When she realized as a teen that she wanted to become a physicist, the desire seemed crazy—and not just because she was a woman. The field was considered dead. At one point, Germany's head of national standards declared, "Nothing else has to be done in physics than just make better measurements." There simply weren't any jobs for physicists. So to want to go into the field meant being driven by passion alone. A life without physics was not a life. Meitner persevered.

Meitner formed a long-lasting research partnership with Otto Hahn after she arrived in Berlin in 1907. She'd recently completed a doctorate in physics in Vienna—Austrian universities had finally started allowing in women—and she was interested in the work Marie Curie was doing in radioactivity. Curie, however, declined to take on Meitner. Max Planck had reservations about allowing women into higher education, but he agreed to let Meitner study and attend lectures, though she already had her PhD, to deepen her understanding of physics. (Meitner grew to adore Planck; Planck, likewise, became one of Meitner's greatest advocates.) Eager to do more than just attend lectures, Meitner paired up with Hahn, a chemist in search of a physicist collaborator.

Hitler forced their separation in the end, but in the beginning, their partnership was divided by the facility's regulations, which did not admit women. Meitner was permitted to conduct radiation experiments in the same chemistry institute as Hahn, but only from a damp, converted carpentry shop in the basement, accessible by a separate entrance. If she needed to use the facilities, she had to walk to the hotel down the street.

Meitner needed to get up to speed on radiochemistry, and that was tough when much of the action was taking place upstairs, but she worked through it. Their first year together, Meitner and Hahn published three articles in radiochemistry. And in 1908, thanks to the Prussian decision that allowed women to enter university, Meitner was finally permitted to use the main building.

Initially shy, Meitner's confidence grew with the strength of her work. In 1912, Planck offered her an assistantship at the university, and Hahn and Meitner relocated to the new Kaiser Wilhelm Institute for Chemistry. By 1917, Meitner ran her own radiophysics department at the institute.

During World War I, both she and Hahn helped out with the war effort, Meitner as an X-ray nurse at an Austrian army hospital, and Hahn as a researcher in poison gas. The pair filled their breaks in service with the quest for an element found in uranium ore, the parent of actinium. They knew the former radioactive element slowly decayed into actinium. With Hahn advising on chemical procedures and Meitner conducting the experiments, in 1917 they lassoed a rare-earth element and named it protactinium.

After their discovery and the conclusion of World War I, Hahn and Meitner remained friends but parted ways as research partners. With radioactivity discovered in 1896 and the splitting of the atom's nucleus in 1917, the field of physics cracked open. All of a sudden Meitner was practicing in a golden age. Chemistry, Hahn's specialty, didn't have the same fire.

In 1934, after many years apart, Meitner convinced Hahn to team up again, to join the race against Enrico Fermi, Ernest Rutherford, and Irène Joliot-Curie to track down new heavy elements. He agreed, and they began sending neutrons speeding toward the atoms of uranium, the heaviest naturally occurring element. They believed that by staging these collisions, they were creating new, even heavier man-made elements, so-called transuranics. But what they were doing was actually something much, much bigger: nuclear fission. None of them—not Curie, not Fermi, not Hahn, not Meitner—knew it. Meitner and Hahn had gotten close to the fission discovery a few times, but in thinking they were after something else entirely, they missed it by a smidge.

Hitler's ethnic cleansing mucked everything up. Meitner was forced to flee, and so her work and the team were left to operate without her physical presence. She and Hahn exchanged letters every other day, but their collaboration had to stay secret. Meitner landed in Sweden, where she found herself stifled by institutional bureaucracy. She had neither the scientific equipment nor the support to carry out more studies, and the distance between her location and the work she loved was frustrating.

Meanwhile, Hahn continued smashing particles. When he found a curious result in 1938, he wrote Meitner asking for help: "Perhaps you can propose some fantastic explanation." Hahn noticed that the collisions were creating not massive heavy elements but smaller ones, roughly half the size of uranium.

During a cross-country ski trip with her nephew, also a physicist, Meitner mulled over the results. Talking through the problem, she and her nephew started to see the data in a different light. What if uranium's nucleus wasn't solid and stiff like a fruit pit? What if it was more jiggly than that, like pizza dough spun in the air? When the center got too thin, the two parts would split, breaking into halves of different proportions—barium and krypton, rubidium and cesium—pairs with combined protons that equaled uranium's 92. And there she had it. They hadn't been making transuranics at all. They'd been creating nuclear fission.

Word of the explanation traveled fast, but as it happens in the game of telephone, parts of the message dropped out. As the good news spread, Meitner, who had both initiated the experiment and explained its outcome, lost credit. The omission was made worse when Hahn claimed full responsibility for the discovery. Furthermore, some scientists believed that a Swedish experimental physicist with powerful sway over his discipline's Nobel Prize may have also blocked Meitner's chances at official recognition. Hahn received the 1944 award without Meitner.

After the war, Meitner was invited to return to her old stomping grounds to become the director of the Max Planck Institute for Chemistry, which had since moved to Mainz. Still deeply affected by the atrocities that had occurred under Hitler, Meitner declined. She felt many of her former colleagues didn't fully comprehend what they'd silently participated in. She couldn't go back.

Though war and circumstance threw her aside for a while, Meitner lived up to the dream she had had as a teenager: "Life need not be easy, provided only it was not empty." Thirty years after her death, a new, man-made, heavy element was named in her honor. Meitnerium is made by fusing bismuth and iron.

IRÈNE JOLIOT-CURIE

1897–1956

CHEMISTRY • FRENCH

WHEN IRÈNE JOLIOT-CURIE WAS SIX YEARS OLD, HER PARENTS, Marie and Pierre Curie, won the Nobel Prize in Physics. Then, when she was fourteen, her mother won a second one, this time in chemistry. The list of Curie Nobel Prize winners got longer when Joliot-Curie herself, at age thirty-seven, shared a Nobel in chemistry with her husband, Frédéric Joliot. "In our family, we are accustomed to glory," Joliot-Curie stated matter-of-factly.

Here "accustomed" is a loaded word, as Joliot-Curie felt the harsh smack of fame from a very early age. The whole world was following along as her parents won their Nobel, but her family was again the focus of attention when Pierre's skull was crushed under a buggy while fumbling with an umbrella. Several years later, when Marie Curie became close with a married colleague, it didn't matter that the details of their involvement were unclear. She was labeled a husband-stealer, which put her status in the scientific community in jeopardy. Joliot-Curie watched her mother accept a second Nobel and then collapse into a full-blown breakdown, sidelining her career for a year and keeping her adoring children from their only living parent. Afraid of the public attention, Joliot-Curie was instructed to send letters to her mother addressed to a pseudonym.

Joliot-Curie both idealized her mother and was extremely protective of her. She shared her mother's love of science, but practiced it with the disposition of her father. Joliot-Curie was calm and confident, where Marie Curie could be fragile and nervous. Once, after being caught daydreaming during a private math lesson, Curie chucked her notebook out the window. Unruffled, Joliot-Curie walked down to the courtyard, picked up the notebook, and was ready with the answer to the math question upon her return. When they were separated when she was a child, Joliot-Curie wrote home with frequent updates, including ones about which equations she found "adorable" (inverse functions) and which one she found the "ugliest" (Taylor's formula). When she grew older, Joliot-Curie made her mother meals, arranged for her transportation, and assisted with whatever else needed to be done.

During World War I, when she was a teenager, Joliot-Curie took up the dangerous task of deploying X-ray technology in field hospitals, following through with a project her mother started. Without X-rays, to find shrapnel in pulverized flesh, physicians had to stick their hands into a wound and dig around. With X-rays and some knowledge of three-dimensional geometry, doctors could evaluate at precisely what angle to enter the wound to retrieve the metal. Joliot-Curie's job was not only to deliver the technology but also to teach hospital staff how to use it. Joliot-Curie's age, sex, and self-possession didn't always endear her to those she tried to teach. In some places she was told not to bother unpacking her equipment, and in others medical workers threatened to destroy the machines as soon as she moved on.

Though she was alone, a teen, and just a few miles from the front, the biggest danger Joliot-Curie faced was actually in the important tools that she helped distribute. With only cotton gloves and a wooden screen to protect her, Joliot-Curie was repeatedly exposed to radiation.

When World War I concluded, Joliot-Curie began working as her mother's assistant at the organization she directed, Paris's Radium Institute. The radioactive glow of materials made her feel giddy. Never one to pick a research subject purely for its popularity, Joliot-Curie let her interests guide her.

Her fluency in physics and math could intimidate her colleagues at times. She didn't care for pleasantries; her verbal and written style struck some as curt and others (like her sister) as simply straightforward and honest. That she was her mother's favorite in the lab earned her the nickname "Crown Princess."

When she presented her PhD dissertation in 1925, even the *New York Times* reported on it: "Nearly a thousand people packed the conference room while the daughter of two of the foremost geniuses of this age calmly read a masterly study." Joliot-Curie, dressed in a loose black dress, explained her analysis of alpha particles cast off by polonium, the element her parents had discovered in 1898. She told one reporter who asked about family obligations, "I consider science to be the paramount interest of my life."

Frédéric Joliot began working at the Radium Institute as Marie Curie's assistant in 1925. He and Joliot-Curie were not at all alike. He was charming and talkative, very perceptive in social situations. Where she ducked the spotlight, he sought it out. But they shared a love for the outdoors and for sports, and a great appreciation of each other's work. Joliot, too, had idolized the Curies. When he was younger, he cut their pictures from a magazine and hung them on his wall. He explained, "I discovered in this girl, whom other people regarded somewhat as a block of ice, an extraordinary person, sensitive and poetic, who in many things gave the impression of being a living replica of what her father had been . . . his good sense, his humility." A year after he began work at the lab, they were married.

Together, Frédéric and Irène swung for a Nobel three times.

In the early 1930s, the Joliot-Curies (as they'll henceforth be called) observed protons flying out of paraffin wax. This is what they knew: the German physicist Walther Bothe had shown how, by placing polonium (which is radioactive) next to beryllium (a brittle metal), the latter will start emitting powerful rays. But what were those rays? The Joliot-Curies guessed gamma rays.

They had misinterpreted their data. When other scientists tried what the Joliot-Curies had done, putting paraffin wax in front of those rays, a subatomic particle with no electric charge appeared: the neutron. (For his discovery of the neutron, James Chadwick won the Nobel Prize in Physics in 1935.)

The Joliot-Curies moved on to studying the neutron in a Wilson cloud chamber. Like following a jet's contrail to track its path, the chamber allowed researchers to study a particle by observing where it had been. The neutron's activity within the chamber could have been explained either by a negatively charged electron or by a positively charged electron, called a positron. They guessed it wasn't a positron and were wrong again.

The Joliot-Curies finally connected with a right answer when they put polonium next to aluminum foil and neutrons and positrons leapt out. The activity was curious because they were expecting to see hydrogen nuclei instead. When they retried the experiment in 1934, they got the same results.

A Geiger counter, which measures ionizing radiation, finally clarified what the Joliot-Curies had done. As the machine clicked, they realized that the aluminum foil had become radioactive. They had discovered the first-ever artificially produced radioactive element. Just months before Marie Curie's death, Joliot brought her a tiny test tube holding the couple's discovery.

In 1935, the Joliot-Curies' artificial radioactivity netted them the Nobel Prize in Chemistry. Following the award, Joliot was hired by the Collège de France and Joliot-Curie stayed on as the director of the Radium Institute. She was also tapped to be one of France's first female cabinet members, though French women were not yet allowed to vote.

As the years went on, health and political problems encroached. During World War II Joliot-Curie and her two children were smuggled out of France. They were fortunate enough to hike the Jura Mountains into Switzerland on June 6, 1944, D-Day, when Germans guarding the Swiss-French border were otherwise occupied. Joliot-Curie made sure to carry a large physics book with her in a backpack.

In 1956, Joliot-Curie was diagnosed with leukemia, likely caused by her exposure to X-rays as a teen. Concurrently, her husband was battling a severe case of hepatitis, also brought on by prolonged exposure to radiation. Joliot-Curie died within the year. Joliot followed two years later. Joliot-Curie was neither terribly surprised nor devastated by the leukemia diagnosis, since the disease had also killed her mother. "I am not afraid of death," she wrote a friend. "I have had such a thrilling life!"

MARIA GOEPPERT MAYER

1906–1972

PHYSICS • GERMAN

MARIA MAYER WAS POISED TO BECOME THE SEVENTH GENERAtion of professors in her family. When she was twenty-three, she left Germany. She thought her chances of snagging a research position at a university were better in the United States. But when Mayer landed in Baltimore, Maryland, in 1930, she was surprised to find that the professional reception was cold.

Mayer had been trained with a group in Göttingen, Germany, that Max Born called "probably the most brilliant gathering of young talent then to be found anywhere." Among her peers were Enrico Fermi and Eugene Wigner. Although some were cowed by their company, she was not one of them. During one quantum mechanics seminar led by Born, Robert Oppenheimer (aka the future father of the atomic bomb) interjected so many questions and comments that Mayer sent around a petition to her fellow students, asking them to sign if they wanted Oppenheimer to keep his trap shut.

Her thesis was completed in a famous theoretical physicist's guest room. One of the bedrooms had a wall signed by Albert Einstein (at the host's invitation, of course). With a legion of physicist friends and supporters, Mayer assumed her transition into American employment wouldn't be too thorny.

But her timing wasn't ideal. Jobs were scarce during the Great Depression. Johns Hopkins University hired her husband as an assistant professor, but because of antinepotism rules, though she was allowed to work, she couldn't be paid for it. Refused a vacant office, Mayer set up a research space for herself in a university attic.

She reported to her little corner of the school every day, not out of obligation, but because she loved physics and was confident that she was a meaningful contributor. Though she was never paid, Mayer was eventually given classes to teach and an office in the physics department. In her nine years at Johns Hopkins, she published ten papers and coauthored with her husband, Joe Mayer, a textbook used in schools for over four decades.

Joe Mayer lost his job in 1938 in a round of firings aimed at restructuring the budget. He landed on his feet—and with a better salary—at Columbia University. For Maria, relocation was more of a blow. Despite her impressive credentials, those who didn't know her often dismissed her as just a professor's wife. Outgoing and confident when she lived in Germany, in the United States Mayer was often shy in her new environment. When she taught classes, she relied on cigarettes to steel her nerves. (Sometimes she'd puff on two or more at a time, lighting up a second while finishing the first.)

Mayer inquired about a position in the Columbia physics department, but her request was spiked. Maria eventually found an office and a bit of support from the chemistry department. There she was given some classes to teach and a job title, although these crumbs were handed to her as if they were favors instead of the embarrassing miscalculations of her value that they were.

Mayer was finally paid for her research during World War II, although it was actually the government, not Columbia, footing the bill. Mayer oversaw a team of some fifteen chemists working on projects related to enriching uranium. After the war a flock of famous physicists moved to the University of Chicago, and the Mayers went with them. Again, the pay was bad (read: nonexistent), but Mayer found the science to be cutting edge, competitive, done at a breakneck pace—in a word, delicious.

When Mayer arrived, a former student offered her an appointment in nuclear physics at Argonne National Laboratory, which would allow her to continue teaching at the University of Chicago. "But I don't know anything about nuclear physics," she pointed out. His reply: "Maria, you'll learn."

So half-time, Mayer wielded a piece of chalk in one hand and a cigarette in the other, running a brutally difficult physics theory seminar and participating in other faculty-like responsibilities at the University of Chicago in exchange for a "voluntary professor" title and no pay. The rest of the time, she worked at Argonne, and puzzled over isotopes.

Isotopes are atoms of the same element that are either carrying extra neutrons or missing a few. Mayer wanted to figure out why some isotopes are more stable than others. She collected evidence showing that the more stable isotopes were the ones with a so-called magic number of neutrons or protons—that is, with 2, 8, 20, 28, 50, 82, or 126. Because these magic-numbered isotopes tended to stay put instead of decaying further like their more unstable counterparts, the universe is loaded with them. Mayer could see that they were special, but she didn't know why.

Like the rings of an onion, Mayer came to believe that inside the nucleus, there were a series of shells. The theory had been first floated by scientists in the 1930s but never proven. Over time, Mayer's data offered strong support for the assumption that neutrons and protons rotated at different levels of orbit. However, she still lacked crucial information that would explain *why* they flew in that formation. Why these numbers and why the onionlike shells?

As she discussed the issue with Enrico Fermi in her office in 1948, he tossed out one last question as he headed out the door: "Is there any evidence of spin-orbit coupling?" Within ten minutes, she had it. His question had brought all her other evidence into focus. The way the particles spun in these onion rings determined the isotope's stability. She compared the model to sets of dance partners waltzing in a room. "Anyone who has danced a fast waltz knows that it is easier to spin in one direction than in the other direction," wrote journalist Sharon Bertsch McGrayne. "Thus, the couples spinning in the easier direction will need slightly less energy." The magic-numbered isotopes were more stable because the waltzing couples were all moving in the direction that produced less energy. Mayer had figured out the nuclear shell model, explaining both what goes on in the nucleus and why some isotopes are more stable than others.

Mayer was offered a full-time university position in California in 1960. (The University of Chicago counteroffered, but after so many years of working there unpaid, she found the gesture amusing.) Three years after she moved to the West Coast, Mayer won the Nobel. She was in bad health then, weakened by a serious stroke and a lifetime of smoking and drinking. Mayer may have slowed down, but she didn't stop. "If you love science," she said, "all you really want is to keep on working."

MARGUERITE PEREY

1909–1975

CHEMISTRY • FRENCH

FROM 1929 TO 1939, MARGUERITE PEREY SLEPT NEXT TO RADIO-active material—she worked with radioactive elements at Marie Curie's Radium Institute and often took her work home. "Home" happened to be a tiny house with bars on the windows, separated from the Radium Institute's lab only by a garden. But while she was there, the space was hers. When she needed time alone, Perey walked her materials across the garden and shut the door. No one else was allowed in. "In those days we took a minimum of precautions," Perey told a reporter decades later. "It was even the thing to scorn dangers of this sort."

Perey started at the Radium Institute in 1929 at the age of twenty. The director of the Institute was two-time Nobel Prize winner Marie Curie. All Perey had was a technician's degree. When she was young, Perey had dreamed about becoming a surgeon, but her family was financially strapped. Her father died when she was four, the family business suffered in a stock market crash, and her mother couldn't afford to finance Perey's education. But what she lacked in formal training, she gained in hands-on education. "Under Marie Curie, I suddenly found myself in the midst of the greatest French chemists," Perey remembered. "And there I was, with only a poor diploma." Perey's curiosity and diligence appealed to Curie, who promoted Perey to be her assistant. (To put Curie's belief in Perey in perspective, Curie once turned down the brilliant scientist Lise Meitner, who'd applied to the institute after completing a physics PhD.)

Perey's first assignment was to prepare sources of the radioactive element actinium for Curie's experiments. Purifying actinium, which was mixed with rare earth materials, was notoriously tricky, not to mention time-consuming, since Perey also had to remove the actinium-related radioactivity that clouded the sample.

Four years in, when she was twenty-four, a sore appeared on Perey's left arm. Because it resembled a burn, Perey's family insisted it was probably just some acid from the lab irritating her skin. Perey had a nagging feeling it wasn't acid doing the damage. A handful of years later, an identical sore surfaced on her right arm. But never mind about that; all this business with actinium was just starting to get interesting.

After performing the same task steadily for a decade, Perey knew exactly what to expect when she pulled a sample of actinium. She handled the task with impressive dexterity, improving the purification steps after years of slogging through them. With her adept handling, she was able to accelerate the process. But in the fall of 1938, when measuring a bit of actinium just recently purified, she detected something she'd never seen before; it appeared to be a new radiation.

A few months later, in January 1939, at the age of twenty-nine, the lab technician pinned down an incredible solution. When Perey followed that surprising radiation to its source, she found a new radioactive element—element 87.

Element 87 filled an empty square in the periodic table's alkali metal group, and it completed the table's spaces for naturally occurring elements. Element 87 had been within scientists' grasp for forty years, but it had escaped detection because no other radiochemist had been fast enough to spot it. Element 87 is both the most rare and the most unstable of all the natural elements. Just 24.5 grams of the stuff are in Earth's crust at any given time With a twenty-two-minute half-life, catching it required Perey's singular speed and skill.

Marie Curie died several years before Perey's discovery, but her daughter Irène Joliot-Curie made sure Perey knew her mother would have been proud. Following Curie's example of honoring her country when naming a new element, Perey anointed element 87 "francium."

Ten years after she took up residence in her little radioactive cave, Perey emerged (and moved out) victorious. Urged by Joliot-Curie to complete her university education, the radiochemist took classes at the Sorbonne during World War II, finally earning her PhD in chemistry in 1946. She stayed at the Radium Institute for twenty years, climbing the ranks from personal assistant to radiochemist to head of research for the National Center for Scientific Research. In 1949, with Joliot-Curie's blessing, Perey moved to the University of Strasbourg to become the chair of nuclear chemistry, eventually founding her own radiochemistry lab in the tradition of her longtime employer. The lab grew rapidly, hosting a bustling community of some one hundred scientists, students, staff, and, yes, research technicians.

When Perey turned forty, she was finally given an official diagnosis for the ailment that had haunted her for over fifteen years. Those sores on her arms were a sign of radiodermatitis, acquired from repeat exposure to radioactive materials. Doctors tried to stop the cancer from spreading, but by the age of fifty, Perey had undergone twenty surgeries, lost two fingers to the disease, and was in such poor health that her caretakers discouraged her even from reading. As her health declined, Perey was forced to remove herself from the university and her thriving lab.

Though a tragic side effect to an incredible career, Perey's high-profile health problems helped usher in crucial occupational regulations in 1960 that would protect others from a similar fate.

In 1962, Perey was the first woman elected into the French Academy of Sciences. It was an honor that, fifty years prior, her mentor had so desperately wanted. (Curie's admission had been defeated by one vote, and a scandal over her candidacy ensued.) Perey was sick then, and getting sicker. As she convalesced in Nice, she reflected on a comment her cousin had made about her scientific fame. "You are the second celebrated person in the family," the cousin said. "In the sixteenth century, one of our ancestors also made a name for himself. He was called Martin the Brawler." Finding a new element was decidedly more dangerous.

CHIEN-SHIUNG WU

1912–1997

PHYSICS • CHINESE

TWO MEMBERS OF THE DIVISION OF WAR RESEARCH AT COLUMbia University spent an entire day grilling Chien-Shiung Wu about her work in nuclear physics. Regarding their own top-secret projects, the interviewers remained dutifully mum until the very end of the day, when they asked if Wu had any idea what they were up to. She cracked a smile. "I'm sorry, but if you wanted me not to know what you're doing, you should have cleaned the blackboards." They asked her to start work the next morning.

From a very young age—from birth even—Wu was groomed for great things. Her name means "courageous hero," and her father was an outspoken advocate not only for his daughter, but for women in general. Her father founded the first school for girls in her area, located outside Shanghai, China. But his advocacy could take her only so far. There weren't any graduate-level options for physics in China, so Wu hopped on a boat to the United States in 1936, planning to leave her family only temporarily while she studied at the University of Michigan. But upon landing in San Francisco, Wu considered the University of California, Berkeley. After weighing the opportunities available for women at both institutions and the academic prestige of the physics programs, Wu chose to stay in California. At Berkeley, like everywhere else, she excelled. A couple years after she received her PhD in 1940, Enrico Fermi was told that if he wanted to find an answer to his self-sustaining nuclear fission problem, he should call Wu.

After staying on at Berkeley for two more years, Wu moved to the East Coast. She was hired by the aforementioned Division of War Research at Columbia University during World War II. After the war concluded, Columbia asked her to stay.

Finally settled, Wu turned her focus to Fermi's theory on beta decay. Beta decay is a type of radioactive disintegration. The theory predicted what would happen to particles in the nucleus of some radioactive elements. But as it stood, scientists were seeing something different from what Fermi had predicted.

Wu immersed herself in the existing research before designing the experimentation. According to Wu, taking on a new area of study meant that "you must know the purpose of the research exactly, what you want to get out of it, and what point you want to show." Furthermore, she believed that for an experiment to be accepted by the scientific community, it not only had to be right; it also had to be able to show unequivocally where other experiments had been wrong. She was incredibly meticulous.

Here's how the process went with Fermi's theory: Wu proved others were getting wonky results because their base materials weren't uniform. When the nucleus ejected its electron, the meatier parts of the material slowed the electron down, making it appear as if the evidence didn't match the idea. When Wu tried the same thing with a uniform base material, her experiment's results aligned with Fermi's theory brilliantly.

Research was deeply satisfying for Wu, so much so that even her most dedicated students and colleagues had a hard time relating to just how profoundly physics stimulated her. On her way home from business trips, she'd ask the cabdriver to swing by the lab just so she could glance at the windows. A graduate student remembered getting an excited announcement from her one weekend morning: "Equipment all alone. Nobody working. Equipment all alone." A colleague affectionately called her a slave driver. Once, hoping for a break from their supervisor's attention, Wu's students bought her tickets to a children's movie, intended as a treat for her and her son. Wu sent the nanny instead. Her friend and quantum mechanics pioneer Wolfgang Pauli once commented that she "is as obsessed with physics as I was in my youth. I doubt whether she ever even noticed the light of the full moon outside."

In the 1950s, physicists were finding new subatomic particles at an astonishing rate, thanks to particle accelerators. Tsung-Dao Lee, a physicist at Columbia, and his research partner, Chen Ning Yang at Princeton, were trying to crack the code that was the K-meson, a newly discovered particle. To the researchers, it seemed as if the K-meson's decay patterns defied the laws of physics. Basically, physics said that the goings-on in the nucleus should be symmetrical, so if electrons were jettisoned from one side, an equal number would be jettisoned from the other. The K-meson, however, was strange. It appeared—and it was blasphemy to say so—that K-mesons favored one side or another.

As was often the case when there was a difficult problem to solve, the Columbia researcher went to Wu in 1956 for guidance. Lee wanted to know if anyone had ever confirmed unequivocally that the nucleus is always symmetrical. Wu wanted to make sure that they were on solid footing before running after a phenomenon that had a one-in-a-million chance—her estimate—of being right. Yang and Lee ripped through a thousand-page tome on the subject. Nada. No one had experimental proof.

Wu knew that if she didn't jump on the experiment immediately, someone else would beat her to it.

On the anniversary of Wu and her husband's twentieth year away from China, they'd planned to take a trip back. But all of a sudden, the K-meson research felt so urgent that she was compelled to bow out of the trip. (Her husband went alone.) What was at stake for the project? Oh, only disproving what was supposed to be a fundamental law of physics.

It took her months to map out a plan and test her tools. To pull off the highly complicated experiment, she needed superpowerful magnets (outpowering Earth's magnetic force by a factor of tens of thousands), a facility that could chill elements to near absolute zero (aka rock bottom on the thermodynamic scale), and cerium magnesium nitrate crystals (brewed in a beaker on her graduate student's kitchen stove). For months, Wu functioned on no more than four hours of sleep.

Operating out of the National Bureau of Standards in Washington, DC, in 1957, Wu coaxed the K-meson into an observable state and watched the electrons fly . . . from one end more than from the other. "These are moments of exaltation and ecstasy," Wu told a reporter. "A glimpse of this wonder can be the reward of a lifetime."

When the results were announced, an article in the *New York Post* gushed, "This small modest woman was powerful enough to do what armies can never accomplish: she helped destroy a law of nature."

ROSALYN SUSSMAN YALOW

1921–2011

PHYSICS • AMERICAN

IF ROSALYN YALOW WANTED TO SEE ENRICO FERMI SPEAK, SHE would *see* Enrico Fermi speak—even if it meant hanging from the rafters. One of the world's greatest physicists talking about one of the world's greatest discoveries? Fermi on fission? She would be there even if she, a junior at Hunter College, had to compete for seat space with every physicist within traveling distance. Yalow did attend his colloquium at Columbia University. And she did see it hanging from the rafters.

Such was the way with Rosalyn Yalow. Once an idea had settled, the obstacles didn't stand a chance. How does a child get braces if her parents are poor? Yalow folded collars with her mother to bring in the necessary cash. How does a researcher secure lab space when she isn't given any? Yalow fashioned one of the first in the United States dedicated to radioisotope research out of a janitor's closet. How does one get past discrimination? "Personally," explained Yalow, "I have not been terribly bothered by it . . . if I wasn't going to do it one way, I'd manage to do it another way." That principle is how she navigated so many issues—graduate school rejections, work limits on pregnant women, rejection from a major journal, and, yes, a packed house for Enrico Fermi. She simply found another way—and quickly. Whining was a waste of time. Minutes were things she didn't like to lose.

Yalow was direct. She questioned colleagues at conferences and spoke up at meetings. At times, people found her manner abrasive, but Yalow sensed a double standard.

She and her longtime research partner, Solomon A. Berson, dealt in directness. Over the course of their twenty-two years working together, communication turned into what onlookers described as a sort of "eerie extrasensory perception." Their rapid-fire conversations about work would spill out into academic events, over dinners, and into walks around campus. At parties, Berson would have to remind Yalow to curb the shop talk and chat with other people.

They began working together at Bronx Veterans Administration Hospital in 1950. Yalow had landed there three years earlier as a consultant via a full-time teaching position at Hunter College, and before that a position at the Federal Telecommunications Laboratory. Yalow wanted to do nuclear physics and neither Hunter nor the Telecommunications Lab was getting her any closer to that goal. Although it was in the closet, at the VA hospital, she finally had her own lab. Berson was a resident physician, and Yalow brought him on board.

The pair clicked immediately. For eighty hours each week, they worked furiously on iodine metabolism, on the role of radioisotopes in blood volume determination, and on insulin research. With test tubes flying and chemical assays to prepare, there was never an extra moment to waste.

One of their first challenges as a team was to figure out how long insulin injected into a diabetic's body stays there. Yalow and Berson attached a radioactive tag to the insulin in order to monitor how long it stuck around. Through frequent blood sampling, they got their answer: too long. The result was surprising because it meant the insulin was being held up by antibodies when the popular assumption was that insulin molecules were so small that they were able to slip past the body's alarm system.

But why did the body attack the injected insulin? Yalow and Berson traced the problem back to an incompatibility between the human body and the hormone injected, which in the 1950s came from pigs and cattle. Even though the difference between human and bovine insulin was slight, antibodies detected that the insulin was foreign and went after it. The pair's discovery overturned a long-held belief and provided a crucial piece of information for doctors treating diabetics. (Today, insulin is synthetically made to exact human specifications to avoid this problem.)

The greatest takeaway from the experiment, however, was not about insulin at all but about how they learned about it. Over the course of conducting their insulin research, Yalow and Berson measured the antibodies generated as a result of the hormone. Flip that relationship around and what do you get? They'd inadvertently developed a way to measure hormones in a test tube by looking at their antibodies. This process didn't require injecting radioactive material into the body and it was capable of surprising accuracy. They called their technique RIA, or radioimmunoassay.

Together, Yalow and Berson tore through hormone research, interpreting their discovery of RIA as a starting gun. What they learned allowed researchers to tell the difference between patients with type 1 and type 2 diabetes; which children would be able to benefit from human growth hormone treatments; whether ulcers should be operated on or handled with medication; which newborns needed a medical intervention for underactive thyroid . . . and the list goes on. Though others were slow to catch on, within a decade, the RIA technique energized scientists, transforming endocrinology into the "it" specialty in medical research. For eighteen years, Yalow and Berson knocked out hormone after hormone, furiously preparing solutions and loading 2,000 to 3,000 test tubes in twenty-four-hour stretches.

By the time Berson moved on to City University of New York in 1968, much of RIA-related research had already been worked out. Even so, Berson and Yalow reunited on Tuesdays and Thursdays to pull all-nighters at the lab.

In a terrible one-two punch, Berson was hit with a small stroke in March 1972 and then a heart attack during a scientific conference in Atlantic City one month later. The heart attack killed him.

Berson and Yalow were so close that their relationship was nearly familial, and his death hit Yalow extraordinarily hard. Besides losing her friend and research partner, she was concerned about losing her status. For their entire partnership, his had been their outward-facing image. Yalow was devastated by his death, but she also didn't want the public interest in her work to be buried with Berson.

Yalow thought that going back to school for an MD might give her extra clout, but with so many years of important research already under her belt, she decided against it. Making a name for herself—by herself—would require upending more than twenty years of assumptions that Berson led the partnership. (Yalow and Berson had always considered each other equals.)

The only way to regain the scientific community's trust, Yalow decided, would be to kick her already breakneck pace up a notch. She turned eighty-hour workweeks into hundred-hour ones. She renamed her lab the Solomon A. Berson Research Laboratory so that her articles—sixty produced in the following four years—would still appear with his name on them.

Yalow knew that her work with Berson deserved a Nobel, but science's highest award is given only to the living, and her partner was already gone. Yalow, as always, didn't give up hope. Every year she chilled champagne and dressed up the day the awards were announced, just in case the news was good.

In the fall of 1977, Yalow woke up in the middle of the night, no longer able to sleep. As was her tradition, if sleep wasn't panning out, she'd go to the office. On this particular morning she was in by 6:45. When she got word that she'd won, Yalow ran home, changed her clothes, and was back in her lab by eight a.m. Her Nobel was granted, but as an exception; both members of a research partnership should have been living.

The Nobel finally affirmed a desire she'd had since the age of eight: to become a "big deal" scientist. This time her admittance was granted with flung-open doors, not a spot in the rafters.

EARTH AND STARS

MARIA MITCHELL

1818–1889

ASTRONOMY • AMERICAN

MARIA MITCHELL WORKED AS A LIBRARIAN BY DAY, BUT IT WAS her other office—a makeshift observatory on the roof of her parents' home in Nantucket, Massachusetts—that was her favorite workspace. She worked there amid spiders and bugs and a stray cat, on both frigid nights and warm ones, studying the stars. "One gets attached (if the term may be used) to certain midnight apparitions," Mitchell wrote in her diary in 1854.

Mitchell started "sweeping the heavens" with her telescope as a child. Her father, who was an astronomer and teacher, was fond of dragging his ten kids upstairs in the evening to stargaze. For her siblings, it was a familial obligation. For Mitchell, it would become her life's work. On October 1, 1847, as she'd done many times before, Mitchell snuck upstairs while her family entertained guests—and this time she caught an extraordinary show. With a puny two-inch telescope (an indication of her skill as an astronomer), she spotted a smudge not visible with the naked eye. When she ran down to tell her father she'd caught a comet in her sight, he wanted to announce it immediately. But Mitchell was cautious. Before taking credit for anything, she wanted to observe the streak more closely to make absolutely sure she had it.

At age twenty-nine Mitchell became among the first Americans to discover a comet and chart its orbit. Her achievement made international headlines, and Mitchell became an instant scientific celebrity. The comet was named "Miss Mitchell's Comet." In her honor, she was awarded a gold medal for the achievement by the king of Denmark, and she was voted into the American Academy of Arts and Sciences. "Fellow"—the customary title for male members—was erased and "Honorary Member" replaced it.

"For a few days Science reigns supreme—we are feted and complimented to the top of our bent . . . one does enjoy acting the part of greatness for a while!" Mitchell wrote. But she also reserved an eye roll for what she came to view as an unnatural display. "It is really amusing to find one's self lionized in a city where one has visited quietly for years; to see the doors of fashionable mansions open wide to receive you, which never opened before. I suspect that the whole corps of science laughs in its sleeves at the farce."

The next wave of attention came in the form of job offers. She did observational work for the US Coastal Survey, which paid Mitchell a salary of $300 by 1849. Then, in 1865, Mitchell accepted a position at Vassar College, made more desirable by its access to the university's top-of-the-line twelve-inch telescope. Giving her students hands-on time with the tools of the trade, however, was significantly trickier. The all-female student body had a curfew, which meant astronomy classes were to be held in broad daylight. Astronomy without the night sky? Mitchell was a high-profile hire without any patience for ridiculous regulations.

She swiftly loosened the school's grip on her students, and lobbied for even more immersive activities for them, including stints as the government's official eclipse observers in locations across the country. After Mitchell talked her way into a lecture at Harvard, bulldozing her way past a reluctant professor ("I asked him if I might attend. He said, 'Yes,' but he didn't look happy!"), she sent "her girls" at Vassar off to attend the Harvard professor's lectures, too. For advocating on their behalf and for her egalitarian teaching style, Mitchell's students adored her. "It meant so much to come into daily contact with such a woman!" one pupil wrote. "There is no need of speaking of her ability; the world knows what she was. . . . Perhaps one clue to her influence may be found in her remark to the senior class in astronomy: 'We are women studying together.' "

During her lifelong dedication to the sky, Mitchell observed the moons of Saturn and Jupiter, sunspots, and nebulae; and she was instrumental in inspiring another generation of women to look to the sky and do the same. After her death, she was immortalized in the starry landscape she so adored: a crater on the moon was given her name.

ANNIE JUMP CANNON

1863–1941

ASTRONOMY • AMERICAN

THE HUMAN EYE SHOULD BE ABLE TO SEE ABOUT EIGHT THOUSand stars dappling the night sky, location and weather conditions permitting. Now observe this: over the course of her career, astronomer Annie Jump Cannon classified fifty times that many—and held the title for more stars collected than any other human long after her death.

With a number like 400,000, she had to have started young. Cannon established her first observatory in her parents' attic. With no trees obstructing the view, Cannon could stargaze through a trapdoor in the roof. Success for the evening was a three-pronged endeavor: First, she had to check the visibility; next, she had to light a candle made of animal fat; finally, she had to flop open a hand-me-down constellation book. Then and only then was she able to truly immerse herself in Delaware's evening sky.

Although her mother, an amateur astronomer, set Cannon upon the skyward path, her father worried over the evening tradition: "Father was more interested in the safety of the house than in the movements of the stars and it was a sigh of relief which he breathed when the evening vigil was over with the house unburned." It wasn't until his daughter attended Wellesley College that her makeshift setup would cause a problem. On another roof but under the same sky, Cannon failed to notice that a lamp she'd placed in a friend's window had started "smoking like a small engine." By the time she got to it, the room had reached the point of near-total char. Cannon called off her observations and spent the remainder of the evening scrubbing down the furniture and walls, which ultimately had to be repapered.

Pyrotechnic issues aside, Cannon's time at Wellesley only reaffirmed her devotion to the cosmos. In 1896, after earning her master's degree, Cannon started as a research assistant at Harvard College Observatory. Her goal: to capture light from distant stars in order to decode the secrets it contained.

When Cannon claimed stellar spectra as her specialty near the turn of the century, learning about stars by examining their light was a discipline on the rise. It ran parallel to another up-and-coming area of research, looking at what happens to stars over time, how they change from birth to middle age to extinction. Because a single star's life cycle takes place over too great a period for humans to observe, astronomers began gathering portraits of stars at all different life stages. With enough collected data, patterns revealing the secrets of stars as they age would emerge.

To the naked eye, the light from a star is white. But send it through a prism, which separates the beam into its constituent colors, and that spectrum becomes its signature, revealing clues about its temperature, gases, and metals. Cannon captured these unique identifiers on photo plates and analyzed them.

Once Cannon got started, she became a star-classifying machine. As she gathered more and more stellar spectra, other scientists looked for ways to mine it.

Cannon's classifications were delivered in two big chunks. The first was as a part of the *Henry Draper Catalogue*, a nine-volume star encyclopedia that came out in installments from 1918 to 1925 and included spectral data for 225,300 stars, most all of them logged by Cannon.

Her follow-up effort, a part of the so-called *Henry Draper Extension*, included stars harder to spot. The first catalog included stars to magnitude 9. Where magnitude is concerned, the lower the number, the brighter the light. So the first release included stars that could be seen with today's binoculars. The

Extension included stars even harder to view, ones classified all the way down to a magnitude of 11. When the *Extension* was published in 1949, it brought the publication's total star count to 359,083.

Cannon spent much of her time unpacking the data in photos with a magnifying glass similar to a watchmaker's, dictating her notes to an assistant. She didn't invent spectral analysis, but over the years, she certainly streamlined it. Fourteen years after Cannon began work at the Harvard Observatory, her modified classification system became the world standard, still used in a refined form today. Cannon's work, the foundation of so many others', heaved the field of astronomy up from one based on observation to a full-on scientific discipline, complete with theory and philosophy.

For the decades she spent squinting at photos, Cannon was given honorary degrees from Oxford and Groningen and awarded prizes like the National Academy of Sciences' Draper Medal. During her lifetime, Cannon was heralded as one of the greatest women alive. "My success, if you would call it that," she said, "lies in the fact that I have kept at my work all these years. It is not genius, or anything like that, it is merely patience."

INGE LEHMANN

1888–1993

SEISMOLOGY • DANISH

WHEN INGE LEHMANN STARTED STUDYING EARTHQUAKES from Copenhagen, Denmark, in 1925, it was a little bit like someone living in the desert becoming a rain forest expert. Set back from major fault lines and far away from most plate tectonic shifts, Denmark's seismic activity—and all of northern Europe's for that matter—was pretty weak compared to that of the Japans and Californias of the world. Her chosen scientific branch drew little funding; two world wars and the Great Depression would direct money for science elsewhere. It was from this humble platform that "the only Danish seismologist" (or so she called herself) made a major discovery. In 1936, thanks to violent rumblings on the other side of the world, Lehmann discovered Earth's inner core.

At the time, the discovery of Earth's core (what we now know as the outer one) was still fairly new. The seismograph was invented in 1880, and the new technology allowed scientists on one side of the globe to detect seismic waves generated by an earthquake on the other. If Earth's insides were uniform, when an earthquake struck, its seismic waves would radiate through Earth's crust in every direction, traveling outward at a consistent clip like the ribs of a handheld fan. But as scientists started studying the data picked up by seismographs, they noticed that waves generated by a single event might come in at varying speeds or surprising destinations.

In 1914, the German geophysicist Beno Gutenberg came to the conclusion that another layer of Earth—a liquid center roughly the size of Mars—must be responsible for the uneven readings. When seismic waves met Earth's liquid layer, he explained, they refracted, changing course, as light does when it hits glass.

As Lehmann was finishing up university and coming up in the profession, she spent a month studying with Gutenberg in Darmstadt, Germany. "He gave me a great deal of time and invaluable help," she remembered. The visit was part of a longer trip dedicated to installing seismic stations throughout Denmark and getting the stations ready for deployment in Greenland. With the technology still so new, Lehmann's co-workers had "never seen a seismograph before."

In 1928, Lehmann was put in charge of the Royal Danish Geodetic Institute's seismological department. But there was just one catch: most of the time she would be without a staff or anyone to help her with the department's records. Because her post was a one-scientist operation, Lehmann often handled work spillover on the weekends. She'd install herself in her garden, nestled amid oatmeal boxes stuffed with seismograph readings on index cards. Lehmann worked in the sunshine, processing the cards by registering the velocity of seismic waves arriving from across the globe.

As it turned out, Lehmann's unlikely location was a boon to her scientific research. The Danish seismographic network, approximately antipodal to the dramatic plate tectonic activity of the South Pacific, played catcher for the seismic waves generated by massive earthquakes across the globe.

Over time, Lehmann noticed in the data a recurring tic that didn't conform to Gutenberg's core. Some seismograph locations that should have registered waves didn't, while other places documented a signal coming in from an unexpected angle.

By 1936, Lehmann had developed a reputation for being a big stickler for observed data; she didn't have much tolerance for ungrounded theories, so to speak. So she approached the problem by working backward from the hard facts of what she could detect. What Lehmann discovered was that the wonky readings weren't aberrations. Her meticulous work revealed an additional structure: an inner core at the center of Gutenberg's liquid one. The 750-mile-wide solid pinball of metal affected both the speed and the trajectory of seismic waves that passed through it.

It took a long time for Lehmann to gain recognition for her work, but once she did, there was a worldwide pile-on. "Well," she told a colleague in New York, "the way it works is you go along and nothing happens. Then you get one medal and everybody else notices or thinks you're respectable for their medal, so you start to get a lot of medals." Lehmann was able to devote more time to research and academic relationships after her retirement from the Royal Danish Geodetic Institute in 1953 (with some mountaineering and skiing trips slotted in more regularly). By the time she began the second phase of her career, she was one of the most respected seismologists in the world. Her esteem was well earned, and an improvement over Lehmann's early days in research, when a basic level of respect was not always the norm. When she was a child, Lehmann had attended a grade school run by Niels Bohr's aunt. (Bohr won the Nobel Prize in Physics in 1922.) And get this: the teachers treated all the students as equals. "No difference between the intellect of boys and girls was recognized," explained Lehmann. "A fact that brought some disappointments later in life when I had to recognize that this was not the general attitude."

During her so-called retirement, Lehmann hopped around research centers. Serious and always self-sufficient, when Lehmann accepted an invitation to visit the Lamont Geological Observatory in Palisades, New York, for a few months in 1952, colleagues recalled that she insisted on walking from place to place, despite offers of a ride. (Lehmann was in her mid-sixties at this point and without a car.) A young Lamont scientist with a Vespa scooter and a secret mission to get "every prominent seismologist in the world on the back," offered Lehmann a ride. At first she declined. But as she watched him zip around with other colleagues, she decided she had enough observable data to hop on.

In the 1960s, funding finally came in for seismology. During the Cold War, the United States wanted to have its ears tuned to underground nuclear explosions, so it updated its old, outdated system. With these new tools, Lehmann learned more about Earth's insides. When she endorsed another scientist's study, her name provided it immediate credibility.

Even in retirement, she continued to make waves.

MARIE THARP

1920–2006

CARTOGRAPHY • AMERICAN

AT FIRST, THE OCEAN WAS A MYSTERY, WITH DEPTHS UNREACHable by humans. Fishermen imagined that there was no bottom. Later, and up until 1851, the seafloor was thought flat, a smooth basin that swooped from continent to continent, slowly filling with sediment at its edges until the bottom gained enough ground that it emerged from the saltwater. In the mid-nineteenth century, the ocean was imagined as a "great sea-gash . . . a scene the most rugged, grand, and imposing. The very ribs of the solid earth."

By 1910, the idea that continents were once connected was raised and quickly buried when one of the world's most prominent geologists called the theory "poppycock," and everyone else thinking about the mysteries of the ocean floor fell into line. Everyone except Marie Tharp, that is. When she brought the possibility to her research partner's attention in 1952, resurrecting the theory of continental drift was "a form of scientific heresy." The suggestion did not go over well. They fought. She insisted. Tharp's colleague, geologist Bruce Heezen, discounted it as "girl talk." The discussion was tabled for a few years.

Tharp had worked with Heezen to map the ocean floor since 1948. Before that, she bounced around. But taking in mountains of data before making a decision always was her way.

Tharp was born in Ypsilanti, Michigan. Her father created soil survey maps for the US Department of Agriculture, and every time he went to a new post, she went with him. The frequent moving meant that Tharp attended some two-dozen schools before she graduated high school. Her father sometimes brought her along on his soil-mapping expeditions. "I guess I had map-making in my blood," admitted Tharp. She earned degrees at three universities: a double major in English and music (with four minors) at Ohio University, a geology degree at the University of Michigan, and a math degree at the University of Tulsa. She had various employment opportunities but none thrilled her. She found herself "bored as hell" at an oil company, unwilling to spend all day squinting though microscopes, and put off by how long it would take to excavate a dinosaur.

To hold her interest, Tharp joked, she'd need "a once-in-the-history-of-the-world opportunity." She would get it before she was thirty. Within two years of being hired by Columbia University, she was working full-time with Heezen to map the ocean floor.

Tharp and Heezen initially worked from a data set that came from the US Navy. During World War II, navy ships were outfitted with instruments that measured echoes. When the ships sent a sonic ping downward, a stylus would move across a piece of paper like a needle on a record. When the ping returned, the stylus would burn a hole in the paper with an electric spark, marking the ocean floor's depth. The recording process happened continuously, providing Heezen and Tharp with the largest data haul of ocean-floor-depth measurements available at the time. The technology had a little snag. When soldiers opened the ship's refrigerator, the electric power shut off—and so did the instrument's ability to measure accurately. "When that happened," said Tharp, "no echo returned and the sounder recorded depths [as] bottomless as the crew's appetite."

By 1952, the research team had collected tens of thousands of measurements—a few drops in the bucket. Even at that number, most of the ocean floor was still uncharted.

Heezen collected the data, and Tharp mapped it. Spread out over drafting tables set up at Columbia University's Lamont Geological Observatory in Palisades, New York, Tharp weaved depth measurements together to create three-dimensional maps. She'd slap a sizable map key over the ocean floor's most obvious bald spots.

Meanwhile, on a parallel project for Bell Laboratories, Heezen hired a fine arts graduate from Boston to plot underwater earthquake epicenters in hopes of getting a better sense of where currents that ripped apart the company's cables originated. Heezen insisted the student use the same scale that Tharp was using for the ocean floor.

Tharp and the earthquake artist switched on the light table. Tharp placed the ocean floor first, and the fine arts grad added his earthquake map on top. Together, the maps revealed something incredible. Like keys on a flute, the earthquakes lined up along the mid-Atlantic ridge.

And there it was: continental drift was real. It would be two more years before Heezen believed her.

In 1959, continental drift got a sizable publicity boost thanks to Jacques Cousteau, who along with almost everybody else wasn't a supporter of the theory. However, Cousteau was curious, so he sailed to the ridge, put a camera on a sled, and dragged it near the seafloor. "He took beautiful movies of big black cliffs in blue water, which he showed at the first International Ocean Congress in New York in 1959," remembered Tharp. "It helped a lot of people believe in our rift valley."

But Heezen was stubborn. He and Tharp had grand fights in which map weights were hurled and trash cans kicked. Their closeness was nearly familial. Against others, they formed a united front.

Although Heezen eventually came around, his supervisor did not. Heezen's supervisor was so furious about the pair's conclusion that he fired Tharp and made sure Heezen, who was tenured, had a wicked hard time carrying out his work.

Tharp was not defined by a drafting table. Even after she

had lost her official position, she worked from home, installing a guard dog named Inky to protect against unfriendly former co-workers. Fortunately, Heezen had enough contacts to continue the ocean floor exploration, and finally, after years of being told she had to stay home, Tharp was cleared to join the project's research vessel.

The partnership between Heezen and Tharp filled in 70 percent of the globe and completely changed the field of geophysics. In her own words, Tharp says, "Establishing the rift valley and the mid-ocean ridge that went all the way around the world for 40,000 miles—that was something important. You could only do that once. You can't find anything bigger than that, at least on this planet."

YVONNE BRILL

1924–2013

ENGINEERING • CANADIAN

IN HIGH SCHOOL YVONNE BRILL WAS TOLD BY HER PHYSICS teacher that women couldn't amount to anything. At the University of Manitoba she was told that the engineering department wouldn't admit women. Later, a colleague told her that she should expect to work twice as hard to receive the same promotion as the men. This, of course, was before she was recognized as one of her generation's most important rocket scientists. Later in life, she'd punctuate each old story with the warm laugh of someone who had always been confident that she could fulfill her own high ambitions. And once Brill let an idea settle in, there was not much anyone could do to dislodge it.

When Brill was four years old, Amelia Earhart became the first woman to fly solo across the Atlantic. To the young Brill, finding freedom through flight looked extraordinary. It was nothing like what she saw growing up in Manitoba, Canada, as the third and youngest child of Belgian immigrant parents who hadn't made it through high school. But no matter. One heroine taking flight was sufficient to prove that there were far-away places to go and extraordinary things to do.

At ten, she passed the University of Manitoba on a streetcar and decided she'd attend. That thing about the university supposedly not admitting women in engineering? Never mind that; she went, and by the time she graduated in mathematics and chemistry at age twenty, Brill was at the top of her class. Soon after, she secured a one-way ticket to Los Angeles. "I didn't really discuss it with my parents," Brill said later in an interview, laughing. "I just went ahead and got all the paperwork together and left."

During the day, Brill worked as a mathematician contributing to the design of the first American satellite at Douglas Aircraft. At night, she pursued a master's in chemistry through the University of Southern California. Brill was believed to be the only woman working in rocket science in the United States in the 1940s. After several years of mathematics, including figuring out the trajectories of different rocket stage sizes using just a slide rule, Brill found that the purely theoretical work at Douglas Aircraft made her restless. She wanted to see her work actually take off, but to do it, she needed to change specialties. Brill considered a career in chemistry, where she had already earned a graduate degree, but ultimately decided against it because of the field's heavy discrimination against women. "There was just no question," she remembered in an interview with the Society of Women Engineers. "Whereas engineering, as an individual of one, they weren't about to make rules to block your progress, because that was too much trouble." She made the switch.

Brill first worked as a chemical engineer in Southern California before moving to the East Coast, where she worked on turbojet engine cycles and chemical manufacturing performance calculations. At the time, electric propulsion systems were, as she called them, "the cat's meow"—both new and one hundred times more powerful than what was then capable with chemical propulsion. But there was still a lot to learn.

Brill started thinking about a particular, crucial moment—the one that happens just as a satellite is injected into orbit. Like a golfer lining up a putt, satellites often need to make little adjustments once they're placed into orbit. The chemical propulsion system at the time was overly complicated, and electric systems needed too much power.

Years earlier, Brill had studied German rockets and had become fascinated with the potential of their chemical propulsion systems. So she began by "looking at the performance and trying to decide what areas of the periodic table one could put emphasis on to get higher performance fuels." Too busy with her day job to devote any on-the-clock hours to a passion project, Brill worked on weekends and late into the night, hunkering down at her kitchen table with pencils, yellow notepads, and a slide rule. Finally confident that she was onto something after examining the ammonia, hydrogen, and nitrogen produced under different conditions, Brill recruited someone with enough skill to check her work. "I never was afraid to risk my job to further ideas that I thought should be adopted, that were good technical ideas, that maybe somebody considered were a little bit far out. But as long as I knew technically I was on the right—or had the confidence that I was technically on the right path, I'd push it." What she discovered was a more fuel-efficient chemical propulsion thruster that helped satellites carry more substantial payloads and remain in orbit for longer periods of time.

Her electrothermal hydrazine thruster was still used in satellites when she died in 2013. It may have been Brill's best-known contribution to rocket science, but it was by no means her only one. Over the course of her career, which took place in the United States and England, Brill worked on the Nova rockets that took America to the moon, the first weather satellite, the first satellite stationed in the upper atmosphere, the Mars Observer, and the engine for the space shuttle. For this work she was awarded the Resnik Challenger Medal by the Society of Women Engineers, the Wyld Propulsion Award from the American Institute of Aeronautics and Astronautics (AIAA), and the National Medal of Technology and Innovation, among others. "She truly represented the best of what American aerospace engineering and system development should be—a pioneering spirit coupled to a clear vision of what the future of an entire area of systems should be, with the ingenuity and genius necessary to make that vision a reality," said AIAA president Mike Griffin in 2013.

With the Society of Women Engineers Brill spent decades both encouraging women to go into math and sciences and encouraging institutions to give female engineers the recognition they deserved. In return, the society gave her access to a network of women all carving out a then-unconventional career.

There's a story Brill liked to tell about a visitor from another company coming into RCA, where she worked at the time, to give a talk. During his presentation, the visitor asked how many propulsion engineers worked at the company. Brill was the only one. Horrified, the visitor explained that his company employed seventy-five. Just then an RCA program manager piped up: "We believe in quality, not quantity."

SALLY RIDE

1951–2012

ASTROPHYSICS • AMERICAN

BEFORE BECOMING THE FIRST AMERICAN WOMAN IN SPACE, Sally Ride got a PhD in astrophysics from Stanford and subjected herself to five years of astronaut training at NASA. Navy test pilots took her on gut-dropping, 600-mile-per-hour flights 39,000 feet in the air. (Her flight instructor called her the best student he'd ever had.) Ride became an expert at maneuvering a 900-pound robotic arm that would be used to pluck satellites from the sky. She became fluent in control panel switches and circuits, getting to know the 1,800 or so on the orbiter's control panel. Ride endured long days of training with singular focus. In another life, Ride would have been a professional tennis player or a Nobel Prize winner—both were within the realm of possibility—but in this one she reached the top of a stack of 8,079 other space program applicants. She was an ambitious scientist and as soon as her 1983 mission was announced, an instant sensation.

Ride landed on magazine covers and into talk-show opening monologues. After years of impossible hurdles thrown up by NASA to prevent women from flying (no, but seriously: the agency's push for women and people of color was summarized as a "near total failure" in 1973 by the person in charge of it), Ride became living proof that gender wasn't the characteristic that would get one booted from the application pool.

She handled space training with ease. Making it through the gauntlet of inane press questions may have been more of a challenge. *Would or wouldn't she wear a bra in zero gravity? Did she cry over her mistakes?* Questions about how her gender would affect the flight were typically answered with some flatly delivered variation of the following: "one thing I probably share with everyone else in the astronaut office is composure." Or, as she reminded one reporter, "Weightlessness is a great equalizer."

Ride was unflappable. Her strategy the morning of her first launch in 1983 was to approach her preparations as if they were mechanical obligations, so as not to get too overwhelmed by the excitement. The astronaut had a knack for tamping down emotion—even if it meant taming the ultimate thrill. When reporters asked, *Why did you want to go into space? What was it like looking back at Earth?* Ride's answers were sometimes flat. *I didn't dream about going to space, I'm not sure why I applied*, and *I can't describe what it was like to look back on Earth.* To say the view from space was not the same as seeing a picture of it was, at least in the beginning, as eloquent as she got.

Ride was better with the concrete, like learning a task or memorizing a text. During her time as an undergraduate at Stanford University, while majoring in both English and physics, Ride and her doubles tennis partner playfully battled each other over who could more seamlessly work obscure Shakespeare quotes into a conversation. As she rose through the academic ranks, Ride recalled her physics advisor saying, "Well! A girl physics major! I've been waiting to see what you'd look like—I haven't seen one for years!" As would become a common theme, Ride was the only one.

Ride went up in the space shuttle twice. The image of her floating through the cabin, snagging a bag of cashews with a halo of curly brown hair, inspired tremendous hope for girls dreaming of going into science. Her presence brought some admirers to tears and compelled others to act.

When an advertisement for college financial aid services displayed a boy in an astronaut outfit dreaming about his future, Ride's father sent the company a strongly worded letter complaining about the "unconscious (I assume) bias we have in education. . . . As a parent of the first US woman astronaut, I know firsthand that girls also aspire to math and science and we should encourage her to 'get America's future off the ground.'"

Ride was more than just the first woman in space. She served as an essential voice of reason—twice—when NASA most needed it.

On January 28, 1986, Space Shuttle Challenger exploded seventy-three seconds after liftoff. Seven of Ride's colleagues died in the accident. Spaceflight up until then had always been the stuff of dreams. But NASA's push for rapid-fire missions at the expense of safety sacrificed lives. The agency needed to figure out what went wrong and how it could recover.

Of the thirteen people brought in to sit on a presidential commission to review the accident, Ride was the only current NASA representative. She was also responsible for gathering some of the most shocking information regarding the agency's missteps. Ride helped hold her employer accountable. The report concluded that NASA had forced through too many flights, ignored warnings that weather conditions might put astronauts in danger, and was entirely too cavalier about sending humans into space. The Nobel Prize–winning physicist Richard Feynman, who was also a member of the panel, claimed that NASA's jam-packed flight schedule was akin to playing Russian roulette. Ride told a reporter that she wouldn't feel safe getting on another flight right away.

The explosion grounded the shuttle program for two years while NASA regrouped. With more rigorous safety measures put in place, the organization needed to map out a plan to win back the public's trust, while also making important decisions about the kinds of missions that would take the agency forward. NASA put Ride in charge of coming up with a refreshed list of mission recommendations.

For a year, Ride tapped young NASA employees to brainstorm the agency's next move. In her final report, she weighed four recommendations: sending humans to Mars, exploring the solar system, creating a space station on the moon, and the one she was most passionate about, organizing a mission to Planet Earth. Internally, the organization favored big projects that ignited the imagination. The longtime NASA heavyweights guard wanted a mission to Mars; Ride argued for an approach more beneficial to the planet. Mission to Planet Earth's goal was to use space technology to understand Earth as a total system, to learn how man-made and natural shifts affect the environment. "This initiative," she wrote, "directly addresses the problems that will be facing humanity in the coming decades, and its continuous scientific return will produce results which are of major significance to all the residents of the planet." At a meeting of the Senate Committee on Commerce, Science, and Transportation, a senator asked Ride to prove how her preferred mission would be more than just "a better weather report." Following the meeting's conclusion, the same senator gushed that the initiative was "the most challenging and exciting concept that this committee has seen in quite some time."

Finally, Ride had an answer to those questions about seeing Earth from space. The astrophysicist in her saw a fragile planet. Her greatest legacy is convincing NASA that Earth is worth trying to protect.

MATH AND TECHNOLOGY

MARIA GAETANA AGNESI

1718–1799

MATHEMATICS • ITALIAN

MARIA GAETANA AGNESI WAS A CHILD PRODIGY. WHEN VISITing scholars dropped by her home in Milan, Italy, Agnesi's father trotted her out to entertain them. She was expected to recite long speeches from memory in Latin or participate in discussions about philosophy or science with men who made it their life's work. Brought up on Cicero's letters, Virgil's poetry, and books like *How to Learn Latin Quickly,* Agnesi was the oldest of twenty-one children, and was often called upon to help her father climb the social ranks. Her younger sister, a brilliant harpsichordist and composer, was also tapped to impress visitors with her extraordinary abilities.

When Agnesi was twenty-one, she realized that her participation in these showcases was not, strictly speaking, compulsory. She broke the news to her father that she had other plans for her future. Agnesi wanted to enter a convent. The announcement came on the heels of a grand display of her scholarly ambition. In a two-hundred-point document, Agnesi listed the theses she would be capable of publicly defending, in addition to all the ones she already had. However, Agnesi was shy and tired of flashing her intellect for her father's social gain. She wanted to give herself over to God.

Agnesi's father wasn't thrilled with the idea. She possessed an extraordinary brain, and he preferred that she used it. Father and daughter struck a deal. If she agreed to continue her mathematics research, Agnesi could do as much charity work as she pleased from home. The public performances could also stop.

A late-blooming mathematician, Agnesi only began seriously studying the subject in her late teens. As with so many other academic pursuits, she took to it immediately. She studied amid globes and mathematical instruments, plowing through calculus before anyone else in Milan was studying it.

Perhaps Agnesi began her next project as a way to pass her knowledge on to her younger siblings. Or maybe she realized how annoying it was to have mathematics instruction siloed into individual branches and one-off books so that getting an education required hunting down a whole collection of resources and hiring a tutor to fill in the gaps. Whatever the case, Agnesi saw a need for a unified textbook covering algebra, geometry, and calculus, so she wrote one.

As was Agnesi's style, when she decided to take on a project, she went big. In 1748, Agnesi published a two-volume, 1,020-page text called *Instituzioni Analitiche*, believed to be the first mathematics book published by a woman. Agnesi had a printing press brought into her father's home so she could oversee the book's typesetting and verify that her formulas were accurately represented. If a particularly unwieldy equation ran past the bottom of the page, it was printed on a long sheet of paper that was folded up and tucked into the regular-size pages.

Agnesi wrote the book in Tuscan, the dialect that would become modern Italian, instead of her own Milanese. Because she chose Italian over Latin—the language of scholars and one she knew well—it appears the text was aimed at a school-age population from the very beginning. *Instituzioni Analitiche* would provide generations of Italian students with a solid and well-rounded mathematics education.

In England, John Colson, a professor at Cambridge, heard about the book and the impact it was making abroad, and felt that British students urgently needed access to the same information. Colson was getting on in age, so he scrambled to bone up on his Italian in order to translate Agnesi's text. He hadn't yet published the translated manuscript when he died in 1760. The work was finally released in 1801 in English, thanks to a vicar who edited and shepherded it through the publication process.

More than 250 years later, Agnesi's name continues to appear in calculus textbooks: she lends it to a curve that rolls over a sphere like a gentle hill. She wasn't the first to discover the curve, although it was assumed she was at the time; mathematics historians found someone who had claimed it earlier. The "witch of Agnesi," as the curve is called, is actually the product of a mistranslation. In *Instituzioni Analitiche*, Agnesi calls her cubic curve *versiera*, which meant "turning in every direction." Colson translated it as *versicra*, or witch.

When the text was first published, Agnesi received many accolades, including a diamond ring and jewel-encrusted box from Empress Maria Theresa, to whom the book was dedicated. A regular correspondent of Agnesi, Pope Benedict XIV recommended her for a position as a professor at the University of Bologna. She declined.

In 1752, when Agnesi was thirty-four, her father died and she was finally able to claim her freedom. She gave up mathematics and her other scholarly pursuits in order to spend the rest of her life serving the poor, donating her entire inheritance to the cause. Agnesi passed away in 1799 in one of the poorhouses she had directed.

Because of her mathematical contributions and the decades she'd spent providing for others, her hometown once lobbied for her sainthood. Her greatest legacy, however, is the witch.

ADA LOVELACE

1815–1852

MATHEMATICS • BRITISH

ADA LOVELACE (NÉE AUGUSTA BYRON) WAS GIVEN A FAMOUS name before she made her own. Her father was Lord Byron, the bad boy of English Romantic poetry, whose epic mood swings could be topped only by his string of scandalous affairs—affairs with women, men, and his half sister. Little Lovelace's mother had had enough. One month after the girl was born, she took the baby and quit the marriage. Lord Byron left England and never returned.

However brief their time in each other's company, Lord Byron was ever present in Lovelace's upbringing—as a model of what not to be. Worried that Ada might lean toward the lyrical, her mother pushed a practical curriculum of grammar, arithmetic, and spelling on the child. When Lovelace became sick with the measles, she was bedridden, only permitted to rise to a sitting position thirty minutes a day. Any impulsive behavior was systematically ironed out.

It may have been a strict upbringing, but Lovelace's mother did provide her daughter with a solid education—one that would pay off when Lovelace was introduced to the mathematician Charles Babbage. The meeting occurred in the middle of her "season" in London, that time when noblewomen of a certain age were paraded around to attract potential suitors. Babbage was forty-one when he made Lovelace's acquaintance in 1833. They hit it off. And then he extended the same offer to her that he had to so many: come by to see my Difference Engine.

Babbage's Difference Engine was a two-ton, hand-cranked calculator with four thousand separate parts designed to expedite time-consuming mathematical tasks. Lovelace was immediately drawn to the machine and its creator. She would find a way to work with Babbage. She *would.*

Her first attempt was in the context of education. Lovelace wanted tutoring in math, and in 1839, she asked Babbage to take her on as his student. The two corresponded, but Babbage didn't bite. He was too busy with his own projects. He was, after all, dreaming up machines capable of streamlining industry, automating manual processes, and freeing up workers tied to mindless tasks.

Lovelace's mother may have tried to purge her of her father's influence, but as she reached adulthood, her Byron side started to emerge. Lovelace experienced stretches of depression and then fits of elation. She would fly between frenzied hours of harp practice to the concentrated study of biquadratic equations. Over time, she shook off the behavioral constraints imposed by her mother, and gave herself over to whatever pleased her. All the while, she produced a steady stream of letters. A playfulness emerged. To Babbage, she signed her letters, "Your Fairy."

Meanwhile, Babbage began spreading the word of his Analytical Engine, another project of his—a programmable beast of a machine, rigged with thousands of stacked and rotating cogwheels. It was just theoretical, but the plans for it were to far exceed the capabilities of any existing calculators, including Babbage's own Difference Engine. In a series of lectures delivered to an audience of prominent philosophers and scientists in Turin, Italy, Babbage unveiled his visionary idea. He convinced an Italian engineer in attendance to document the talks. In 1842, the resulting article came out in a Swiss journal, published in French.

A decade since their first meeting, Lovelace remained a believer in Babbage's ideas. With this Swiss publication, she saw her opening to offer support. Babbage's Analytical Engine deserved a massive audience, and Lovelace knew she could get it in front of more eyeballs by translating the article into English.

Lovelace's next step was her most significant. She took the base text from the article—some eight thousand words—and annotated it, gracefully comparing the Analytical Engine to its antecedents and explaining its place in the future. If other machines could calculate, reflecting the intelligence of their owners, the Analytical Engine would amplify its owner's knowledge, able to store data and programs that could process it. Lovelace pointed out that getting the most out of the Analytical Engine meant designing instructions tailored to the owner's interests. Programming the thing would go a long way. She also saw the possibility for it to process more than numbers, suggesting "the engine might compose elaborate and scientific pieces of music of any degree of complexity or extent."

Reining in easily excitable imaginations, Lovelace also explained the Engine's limitations ("It can follow analysis; but it has no power of anticipating any analytical relations or truths") and illustrated its strengths ("the Analytical Engine weaves algebraical patterns just as the Jacquard-loom weaves flowers and leaves").

The most extraordinary of her annotations was Lovelace's so-called Note G. In it, she explained how a punch-card-based algorithm could return a scrolling sequence of special rational numbers, called Bernoulli numbers. Lovelace's explanation of how to tell the machine to return Bernoulli numbers is considered the world's first computer program. What began as a simple translation, as one Babbage scholar points out, became "the most important paper in the history of digital computing before modern times."

Babbage corresponded with Lovelace throughout the annotation process. Lovelace sent Babbage her commentary for feedback, and where she needed help and clarification, he offered it. Scholars differ on the degree of influence they believe Babbage had on Lovelace's notes. Some believe that his mind was behind her words. Others, like journalist Suw Charman-Anderson, call her "[not] the first woman [computer programmer]. The first person."

Lovelace guarded her work, and sometimes fiercely. To one of Babbage's edits, she replied firmly, "I am much annoyed at your having altered my Note . . . I cannot endure another person to meddle with my sentences." She also possessed a strong confidence in the range of her own abilities. In one letter, she confided, "That brain of mine is something more than merely mortal. . . . Before ten years are out, the Devil's in it if I haven't sucked out some of the lifeblood from the mysteries of the universe, in a way that no purely mortal lips or brains could do."

For what it's worth, Babbage himself was effusive about her contributions. "All this was impossible for you to know by intuition and the more I read your notes the more surprised I am at them and regret not having earlier explored so rich a vein of the noblest metal."

The Department of Defense named a computer language after her. Ada Lovelace Day celebrates the extraordinary achievements of women in science, technology, engineering, and math. The "Ada Lovelace Edit-a-thon" is an annual event aimed at beefing up online entries for women in science whose accomplishments are unsung or misattributed. When her name is mentioned today, it's more than a tip of the hat; it's a call to arms.

FLORENCE NIGHTINGALE

1820–1910

STATISTICS • BRITISH

ON A COLOR-CODED PAGE LABELED "DIAGRAMS," FLORENCE Nightingale drew a delicate, circular chart. Divided two ways, the chart looked like a dart's target, a series of concentric circles sliced into wedges. The slices were labeled like a clock, but with months replacing numbers, starting with July in the noon spot, August at 1, and so on. The rings each noted a number. The smallest was labeled 100, the second, 200; the outermost circle, 300.

The shaded portion of the chart indicated the number of deaths in British Army hospitals per month from April 1854 to March 1855, during the Crimean War. In July, the light green area (infectious diseases) topped out just above 150. As it got colder, the death toll rose, and a spotlight of green splashed down the page, traveling far beyond the last ring. The chart indicated that the death toll from wounds was fewer than 50. The number of deaths by diseases in the same month: 1,023.

Nightingale's name is synonymous with nursing. She's the lady with the lamp, the compassionate caretaker who checked in on ailing soldiers in the middle of the night. She recognized how atrocious the conditions were in wartime hospitals and lobbied for better standards, based on the needs of patients. It was important work—the foundation of modern nursing—but her statistical analysis of big public health problems is arguably just as influential. In fact, the principles she developed while designing data-gathering tools, and the methods of data analysis and preparation she hammered out, marked the beginning of evidence-based medicine.

When Nightingale was sent to Turkey to support military

hospitals, details about the deplorable conditions had already made their way to the newspapers. Illness was taking down soldiers faster than enemy bullets. What Nightingale did was quantify those sensational stories. The charts she created—what she called "coxcombs," now known as "polar-area diagrams"—were so striking in their visual assessment that when Nightingale began lobbying for change, she had a sturdy platform to stand on. In 1856, Nightingale took her concerns to Queen Victoria and Prince Albert.

It took the British secretary of state for war less than a year from Nightingale's return from Crimea to issue an order that called for the creation of a Statistical Branch of the Army Medical Department. Nightingale's data and her visualizations provided rapid clarity on the failings of military hospitals; improper sanitation was to blame.

After offering her diagnosis, Nightingale laid out a clear set of standards aimed at improving conditions for patients in hospitals. Some recommendations, like installing easy-to-clean walls, floors, and equipment or offering patients food with nutritional value, now seem basic. But ideal qualities, like access to light and quiet, are ones that hospitals still strive for today.

In her book *Notes on Nursing,* Nightingale's most well-known offering, she explained that "the symptoms or the sufferings generally considered to be inevitable or incident to the disease are very often not symptoms of the disease at all, but of something quite different—of the want of fresh air, or of light, or of warmth, or of quiet, or of cleanliness, or of punctuality and care in the administration of diet." Bedsores, for instance, were something that nurses had direct control over. And transferring those burdens from patient to caretaker marked a seismic shift in philosophy.

Through observation and statistical analysis of census data, Nightingale designed a curriculum for nurses that would provide them adequate training for the very first time. The program had its grand unveiling in 1860 at a brand-new school, the Nightingale School of Nursing at St. Thomas' Hospital in London, funded through private donations. Feeling ill, Nightingale wasn't able to attend the opening ceremony.

As Nightingale fought to improve the health of others, she spent more and more time at home trying to protect her own. For decades, Nightingale was plagued with an illness historians now think was brucellosis. During that time, she retreated to her room and rarely left.

Although her poor health eventually halted public appearances, it didn't stop Nightingale from working. She sank her efforts into statistics, which provided a reliable way to suss out patient needs. The better the information, the more effectively Nightingale could go about initiating change. Nightingale also kept up a lively correspondence in letters. By the end of her life, she was writing letters for twelve hours a day, the method she'd long used for keeping up with statisticians, friends, and Nightingale-led efforts in India and Australia to update nursing practices. If she received an inquiry about the proper material for hospital walls, Nightingale would whip out thirteen pages on the intricacies of parian cement. Because letter writing was her primary mode of communication, Nightingale became highly skilled in the art, always present, attentive, and sensitive to her audience.

Becoming a global icon during her lifetime made Nightingale deeply uncomfortable. The focus, she thought, should be on the patient. Though she had long before hung up her lamp, it continued to shine on her.

SOPHIE KOWALEVSKI

1850–1891

MATHEMATICS • RUSSIAN

SOPHIE KOWALEVSKI BELIEVED IT WAS A MISTAKE OF THE UNinformed to confuse mathematics with arithmetic. Arithmetic was just a pile of "dry and arid" numbers to be multiplied and divided. Mathematics was a world of elegant possibilities that "demand[ed] the utmost imagination." To engage in mathematics fully was to elevate it to an art not unlike poetry. "The poet must see more deeply than other people, and the mathematician must do the same."

Looking deeply into the numbers was a skill she acquired at a very young age. When Kowalevski was a child, her father, who had recently retired from Russian military service, moved the family to a rural estate near the Lithuanian border. It was a large home next to a forest and on a lake, far from any big cities. They ordered wallpaper from St. Petersburg to freshen up the home's interior, but when the paper arrived, it became clear that there had been a miscalculation. The nursery was left bare. Instead of going through the hassle of ordering more, Kowalevski's father fashioned an inexpensive, DIY solution. He had the room papered with the lithographed lectures on differential and integral calculus from a course he'd taken as a young officer. If there is an event that catalyzes the imagination, sending us, for the rest of our lives, restlessly after our passions, for Kowalevski, this was it. Her governess could not tear the girl away from the equation-layered room. "I would stand by the wall for hours on end, reading and rereading what was written there." She was too young to understand its meaning, but age didn't stop her from trying.

For the majority of her childhood, Kowalevski's education

did not keep pace with her curiosity. Her father wasn't keen on the idea of "learned women." Consequently, her formal instruction was spotty. "I was in a chronic state of book hunger," she wrote in her autobiography. Kowalevski would sneak into her family's library to consume the forbidden foreign novels and Russian periodicals heaped on the room's tables and couches. "And here, suddenly at my fingertips—such treasure! How could anyone not be tempted."

When her uncles visited, she probed them for stories about math and science. Through them, she learned how a coral reef was formed, how mathematical asymptotes would never kiss the curve leaning toward them, and about the Greek problem of how to square a circle. "The meaning of these concepts I naturally could not yet grasp, but they acted on my imagination, instilling in me a reverence for mathematics as an exalted and mysterious science which opens up to its initiates a new world of wonders, inaccessible to ordinary mortals."

Kowalevski whipped through a borrowed algebra book, ducking the attention of her governess while she studied. When a neighbor, a physics professor, dropped off a textbook he'd written, as a gift for her father, the volume mysteriously ended up in his daughter's possession. The next time the professor visited the house, Kowalevski engaged him in conversation about optics—not the simplest task. The professor was reluctant to talk to her about something that she couldn't possibly understand. She was young—at this point in her teens—and a woman. But Kowalevski's explanation of sine changed his mind.

Because she was mostly self-taught, Kowalevski's education had gaps. The chapter on optics, for instance, gave her trouble because she lacked a foundation in trigonometry that would have explained the function of sine. And sine was all over the place! So she began experimenting with its meaning, ferreting out an answer through trial and error. When she laid out her conclusion for the professor, his jaw hit the floor. She had pioneered her way to sine's meaning via the same route that mathematicians had taken historically.

The professor appealed to her father, comparing Kowalevski's considerable abilities to the famous French mathematician Pascal. She needed advanced academic training, stat.

Her father finally gave in. Kowalevski's opportunities in Russia, however, had a well-established ceiling. Her only chances for greater professional development were abroad. But how to get there? Unmarried, she was stuck at home, subject to her father's rules. Married, she would be forced to conform to her husband's life in Russia. To Kowalevski and her older sister Anyuta, neither option was viable. Kowalevski opted for a third, more unconventional option. She entered into a sham marriage.

Her husband, Vladimir Kowalevski, was part of a radical political group fighting for equal education for women. When Sophie married Vladimir at age eighteen, both she and her sister were free to leave Russia thanks to their new legally bound but platonic chaperone.

Kowalevski's first stop was Heidelberg, Germany. (Her husband went elsewhere to study geology.) But when she arrived, Kowalevski found that women were barred from university enrollment. The young mathematician, though, was practiced at using her insight as a tool to change reluctant minds. Kowalevski soon gained approval to attend lectures unofficially. One classmate, Yulya Lermontova, who became the first Russian woman to earn a doctorate in chemistry, remembered the impression Kowalevski made on the place. "Sofya immediately attracted the attention of her teachers with her uncommon mathematical ability. Professors were ecstatic over their gifted student and spoke about her as an extraordinary phenomenon. Talk of the amazing Russian woman spread through the little town, so that people would often stop in the street to stare at her."

Next, Kowalevski traveled to Berlin, where she convinced a mathematician she greatly admired, named Karl Weierstrass, to teach her privately. (The University of Berlin, where Weierstrass taught, had an even stricter ban on women.) He was no supporter of the other sex in academics, but Kowalevski's abilities and passion for the subject quickly earned her a place as his star student and later a trusted peer.

She wanted a doctorate in mathematics, so Weierstrass facilitated one from the University of Göttingen—a university that would grant higher degrees to women—without Kowalevski having to attend class or exams. From Berlin, Kowalevski became the first woman in Europe to earn a PhD in mathematics. Most doctoral students opted to write one dissertation; Kowalevski assembled three: two in pure mathematics and one in astronomy.

Meanwhile, Kowalevski's sham marriage morphed into a real one. In 1875, she returned with her husband to Russia, putting mathematics aside. Weierstrass begged Kowalevski to come back to Europe and her studies. With so much distance between them, she stopped returning her advisor's letters.

Six years after she left Berlin, having accrued several failed real estate ventures and a strained marriage, Kowalevski returned to Germany alone. Her work resumed immediately. Kowalevski published groundbreaking papers on the refraction of light in crystals and on "the reduction of a certain class of Abelian functions to elliptic functions." In 1883, Stockholm University invited her to become a lecturer. She initially rejected the invitation, citing "deep doubts" about her ability to excel at the position until she felt ready to live up to the honor. However, within six months of her arrival, she'd been promoted to full professor and offered an editor position in the journal *Acta*

Mathematica. Two years later she was the department chair, fluent in Swedish, and dedicated to her work with a singular passion not felt since the early days of liberation from her father's roof.

It was then, egged on by supportive peers, that she went after what the discipline called the "mathematical mermaid," a classical mathematical problem that had eluded many greats. For advancing the field's understanding of this problem, which involved "the rotation of a solid body around a fixed point under the influence of gravitational force," the Paris Academy of Sciences would issue a cash prize. Kowalevski worked furiously to complete her offering on time.

The Paris Academy of Sciences' announcement was a shock for two reasons. First, the winner broke so much new ground on the problem that the prize's governing body voted to increase the pot. The second was only a surprise to those who didn't already know her. Of the fifteen entries submitted anonymously, Kowalevski's took the prize. Her solution led the way to new areas of research in theoretical mathematics. An analysis of her work pointed out that her win had influence that was more than mathematical: "The value . . . is not only in the results themselves nor in the originality of her method, but also in the increased interest she aroused in the problem . . . on the part of researchers in many countries, in particular Russia."

By the time of her death from pneumonia at age forty-one, Kowalevski had risen to the top of her discipline. As was custom, her brain was weighed and assessed, the size and grooves judged as an indication of ability. "[The] brain of the deceased was developed in the highest degree," reported the Stockholm newspapers. "And was rich in convolutions, as might have been predicted, judging by her high intelligence."

EMMY NOETHER

1882–1935

MATHEMATICS • GERMAN

ALBERT EINSTEIN WAS IN OVER HIS HEAD. HE HAD WORKED OUT his general theory of relativity, but he was having problems with the mathematics that would have to correspond. So Einstein pulled in a team of experts from the University of Göttingen to help him formulate the concepts. The team was led by David Hilbert and Felix Klein, who were held in extremely high regard for their contributions to mathematical invariants. But their legacy, in part, is the community of scholars they fostered at Göttingen, who helped the university grow into one of the world's most respected mathematics institutions. They scouted talent. For the Einstein project, Emmy Noether was their draft pick.

Noether had been making a name for herself steadily. In the eight years prior, she worked at the University of Erlangen without a salary or a job title. By the time she left for Göttingen, she had published half a dozen or so papers, lectured abroad, taken on PhD students, and filled in as a lecturer for her father, Max Noether, who was an Erlangen mathematics professor suffering from deteriorating health.

At the time, Noether's specialty was invariants, or the unchangeable elements that remain constant throughout transformations like rotation or reflection. For the general theory of relativity, her knowledge base was crucial. Those interlinked equations that Einstein needed? Noether helped create them. Her formulas were elegant, and her thought process and imagination enlightening. Einstein thought highly of her work, writing, "Frl. Noether is continually advising me in my projects and . . . it is really through her that I have become competent in the subject."

It didn't take long for Noether's closest colleagues to realize that she was a mathematical force, someone of extraordinary value who should be kept around with a faculty position. However, Noether faced sharp opposition. Many of the people who supported the push to make her a lecturer also believed that she was a special case and that, in general, women shouldn't be allowed to teach in universities. The Prussian ministry of religion and education, whose approval the university needed, shut down her appointment: "She won't be allowed to become a lecturer at Göttingen, Frankfurt, or anywhere else."

The shifting political landscape finally cracked open the stubborn set of regulations governing women in academia. When Germany was defeated in World War I, socialists took over and gave women the right to vote. There was still a movement internally to get Noether on staff, and Einstein offered to advocate for her. "On receiving the new work from Fräulein Noether, I again find it a great injustice that she cannot lecture officially," he wrote. Though Noether had been teaching, on paper her classes were David Hilbert's. Finally, Noether was allowed a real position at the university with a title that sounded like fiction. As the "unofficial, extraordinary professor," Emmy Noether would receive no pay. (Her colleagues joked about the title, saying "an extraordinary professor knows nothing ordinary, and an ordinary professor knows nothing extraordinary.") When she finally did receive a salary, she was Göttingen's lowest-paid faculty member.

Pay or no pay, at Göttingen she thrived. Here's how deeply one line of study, now called Noether's theory, influenced physics, according to a physicist quoted in the *New York Times*: "You can make a strong case that her theorem is the backbone on which all of modern physics is built." And the dent she made in mathematics? She was a founder of abstract algebra. In one paper, published in 1921 and titled "Theory of Ideals in Rings," Noether dusted her work free of numbers, formulas, and concrete examples. Instead she compared concepts, which, the science writer Sharon Bertsch McGrayne, explains, "is as if she were describing and comparing the characteristics of buildings—tallness, solidarity, usefulness, size—without ever mentioning buildings themselves." By zooming way, way out, Noether noticed connections between concepts that scientists and mathematicians hadn't previously realized were related, like time and conservation of energy.

Noether would get so excited discussing math that neither a dropped piece of food at lunch nor a tress of hair sprung from her bun would slow her down for a second. She spoke loudly and exuberantly, and like Einstein was interested in appearance only as it related to comfort. Einstein loved his gray cotton sweatshirts when wool ones were the fashion; Noether wore long, loose dresses, and cut her hair short before it was in style. For Einstein, we call these the traits of an absentminded genius. For Noether, there was a double standard—her weight and appearance became the subject of persistent teasing and chatter behind her back. Like the trivial annoyances of title, pay, and politics, the comments didn't bother Noether. When students tried to replace hairpins that had come loose and to straighten her blouse during a break in a particularly passionate lecture, she shooed them away. Hairstyles and clothes would change, but for Noether, math was her invariant.

With a mind working as rapidly as hers, it was a challenge for even Noether to keep up with her own thoughts. As she worked out an idea in front of the class, the blackboard would be filled up and cleared and filled up and cleared in rapid succession. When she got stuck on a new idea, students recalled her hurling the chalk to the floor and stomping on it, particles rising around her like dust at a demolition. Effortlessly, she could redo the problem in a more traditional way.

Both social and generous with sharing ideas, many, many important papers were sparked by Noether's brainpower and published without her byline but with her blessing. In fact, whole chunks of the second edition of the textbook *Modern Algebra* can be traced back to her influence.

Politics in Germany affected her career again. Though Noether had established herself as one of the greatest mathematical minds of the twentieth century, the Nazis judged only her left political leanings and her Jewish ancestry. In May 1933, Noether was one of the first Jewish professors fired at Göttingen. Even in the face of blatant discrimination, perhaps naively, the math came first. When she could no longer teach at the university, Noether tutored students illegally from her modest apartment, including Nazis who showed up in full military gear. It wasn't that she agreed with what was happening, but she brushed it aside for the dedicated student. "Her heart knew no malice," remembered a friend and colleague. "She did not believe in evil—indeed it never entered her mind that it could play a role among men."

For her generosity, Noether's friends were wholly dedicated to her. Understanding that staying in Germany would put her in serious danger, in 1933 her friends arranged for Noether to take a position at Bryn Mawr College in the United States. It was meant to be a temporary post until she could land somewhere more prestigious. But just two years after she arrived, Noether died while recovering from a surgery on an ovarian cyst. She was fifty-three. Following her death, Einstein wrote a letter to the *New York Times.* "Fräulein Noether was the most significant creative mathematical genius thus far produced since the higher education of women began." Today, some scientists believe her contributions, long hidden beneath the bylines and titles of others, outshine even the accomplishments of the ode's writer.

MARY CARTWRIGHT

1900–1998

MATHEMATICS • BRITISH

THE FAMOUS THEORETICAL PHYSICIST FREEMAN DYSON talked himself into the doghouse by singing Mary Cartwright's praises. Cartwright had launched a new field of mathematics, he said. We can attribute chaos theory, which helps explain everything from the weather to the stock market to the way water flows, to her. At age ninety-three, Cartwright wasn't fond of the attention. Yes, she may have come up with the mathematical formulation, but its applications had evolved beyond her. "I heard you were extolling me as one of the pioneers of work on chaos. I do not know what is meant by chaos," Cartwright wrote. "My nephew lent me a large book on chaos and there was no mathematics in it."

Cartwright's chaos theory evolved out of an urgent problem with radar technology. During World War II, radar was considered clutch in the battle against Hitler. But when British soldiers needed more powerful amplifiers, the signal became jumbled. Clearing up the signal could mean the difference between winning the war against Germany and losing it.

In an attempt to fix the system as quickly as possible, in 1938 the British Department of Scientific and Industrial Research appealed to the members of the London Mathematical Society for assistance. According to Cartwright, the government needed "help in solving certain very objectionable looking differential equations occurring in connection with radar." An Oxford-educated lecturer at Girton College at Cambridge with prior experience in hard-to-parse differential equations, Cartwright rose to the challenge, drawing in J. E. Littlewood to serve as research support.

Cartwright first met Littlewood while defending her PhD thesis in 1930. They were drawn together by a wink. When another reviewer interjected with something foolish, throwing Cartwright momentarily off track, Littlewood floated the friendly gesture as a bit of encouragement.

They met again that same year when Cartwright was conducting research on the theory of functions. While working at Girton on a fellowship, Cartwright attended Littlewood's lectures. She had a particular skill for weaving mathematical concepts together in unconventional ways. She came to Littlewood's attention for a second time when she applied a technique used for one kind of problem to the problem Littlewood had posed in class. He published the theorem she designed five years later.

The British government's radar problem wasn't exactly up Cartwright's alley, but it sounded interesting and she knew she could rely on Littlewood to lend his expertise in areas of the research, that were more unfamiliar to her, like radio dynamics. Forged in 1938, their partnership largely took place through the post. From time to time, they'd take a walk and he'd sketch out something with his finger on the fly, but the majority of their interactions were mailed.

Cartwright would work through one section, send it to Littlewood, and await his feedback. On the occasions that she made an obvious error, he scribbled a snake on the page next to the mistake. Littlewood's responses were so often tardy that Cartwright took to giving him gentle reminders whenever she ran into him at the bank or on the street. For a while, they plodded along.

When she began working on the amplifier issue, Cartwright read everything she could get her hands on to ground the problem in existing research. She familiarized herself with the work of the electrician Balthasar van der Pol. Van der Pol's equation was the one scientists and mathematicians turned to when trying to explain a nonlinear amplifier like the one the British government was hoping to improve. While the equation explained a lot, some of van der Pol's outcomes didn't fit the model and he didn't have a good explanation as to why.

Cartwright's first big breakthrough came while she was in the bathtub. During a soak of brilliance, Cartwright combined van der Pol's equation with the work of the mathematician Henri Poincaré. Poincaré studied the movement of celestial bodies by using nonlinear equations. Like van der Pol, he had also formulated a model that contained some irregularities, but this time applied to orbiting satellites. In the late nineteenth century, Poincaré developed a new branch of mathematics to capture these complex trajectories. By applying Poincaré's methods to van der Pol's problem, Cartwright was able to make sense of the irregularities.

It would take years for the pair to work through the solution's implications. In 1945, they finally published their results. The answer to the radar amplification problem is what we now refer to as chaos theory. It's the idea that tiny fluctuations in conditions can produce widely varying outcomes. When it was presented, Dyson remembers, "I saw the beauty of her work but I did not see the importance."

Her model came too late to solve the radar amplification problem in World War II. However, she and Littlewood were able to provide enough information—and get it quickly enough—for the British army to design a work around.

Cartwright's chaos explanation didn't get much play from anyone—mathematicians included—until nearly thirty years later, when the meteorologist Edward Lorenz asked an enchanting question: "Does the Flap of a Butterfly's Wings in Brazil Set Off a Tornado in Texas?"

Cartwright had retired by 1972, the same year Lorenz gave

his famous speech about chaos theory as it relates to weather. By that point, Cartwright had enjoyed a long, impressive career. She was elected president of the London Mathematical Society and given the title "Dame" by the queen of England. When the honor was bestowed in 1969, a colleague joked about having to bow in Cartwright's presence three times. With characteristic dry humor, Cartwright replied, "No. Twice will do."

For nineteen years, she served as Mistress of Girton College at Cambridge, where her responsibilities slowed her research but never stopped it. Teaching gave her a lot of hope for future generations of mathematicians. "Mathematics is a young person's game mainly because major advances in the subject come from approaching problems from a slightly different perspective than previously adopted," she explained. "These types of ideas often come in the course of learning a subject for the first time."

Cartwright helped discover chaos theory early in her career, too. Decades later, she saw chaos theory burst open before her, finding applications in physics, engineering, and meteorology, among others. Though she was never comfortable owning its expansion, her resistance was fitting. One never really can predict where a creative combination of two ideas will end up accelerated decades into the future.

GRACE MURRAY HOPPER

1906–1992

COMPUTER SCIENCE • AMERICAN

EVERY TIME WE CALL A COMPUTER GLITCH A BUG, WE SHOULD give a little nod to the "Grand Lady of Software." Because if it wasn't for Grace Hopper and the moth she found wedged in the hulking Mark II computer's relay, the computer bug might have been known by any other name.

Hopper's influence goes far beyond the bug. Hopper played such a significant part in the early history of computing that her influence, like technology itself, appears everywhere. Her résumé would say she was a computer programmer—and she was—as important to the development of computers as Charles Babbage and Ada Lovelace. But her voice and vision are apparent in both technology and the way we talk about it.

Long before Apple popularized the slogan *Think Different* and being "disruptive" became a Silicon Valley mantra, Hopper lectured students, colleagues, and technology companies against using what she called "the most damaging phrase in the language." What was this cardinal sin of innovation? "*We've always done it this way.*" Hopper was so adamant about banning the phrase, that she, dressed in her full navy uniform, often threatened to—poof!—"come back and haunt" the poor souls who dared to utter the phrase. In any case, the idea has remained a core tenet of technology. Today, the worst thing you can say about a new idea is that it's safe. As a constant reminder to rethink even those things we consider fundamental, Hopper's office clock ticked counterclockwise.

"It's always easier to ask forgiveness than it is to get permission" is another well-known Hopperism—and one she practiced long before fine-tuning its expression. When Hopper was a child, she was powerfully drawn to gadgets. At age seven, she wanted to know how an alarm clock roused her family out of bed each morning. So Hopper took the thing apart. When she couldn't put it back together again, she dismantled another one. Still stumped, she tried another. When she'd pulled the screws and springs from seven machines, Hopper's mother made a deal with the child: she could tinker with one.

Supported by a mathematics-loving mother and an encouraging father, Hopper started at Vassar at age seventeen, earning a degree in mathematics in 1928. From there she went to Yale, knocking out both a master's and a PhD in mathematics (the school's first woman to do so) before returning to Vassar to teach math, the subject she loved.

For Hopper, everything changed when Japan bombed Pearl Harbor in 1941. Hopper, at age thirty-four, wanted to do something tangible for her country; she wanted to enlist. Sure, the government thought that her vocation as a math professor was too important to leave. Sure, she was sixteen pounds underweight and, by average enlistment standards, very old. But Hopper was confident and determined. She wrangled a leave of absence from Vassar, arranged a waiver for her weight, and in December 1943 succeeded in joining the US Naval Reserve.

In the Reserve, Hopper was assigned a post in the Bureau of Ships Computation Project at Harvard University. Her reputation as an excellent mathematician preceded her. As she arrived, her supervisor offered the kind of pleasantries that come after a long, impatient wait: "Where have you been?" He immediately put her to work on the organization's massive Mark I computer, charging her with learning "how to program the beast and to get a program running."

For a mathematician with a gadget obsession, the Mark I—at 51-feet-long and 5 tons—was a dream with a staggering processing rate: some 72 words and three operations calculated every second. Hopper was its lead programmer, its tour guide.

The 561-page manual she wrote for the machine was groundbreaking, according to a computer historian. "The instruction sequences . . . are thus among the earliest examples anywhere of digital computer programs."

After she was released from active duty, Hopper chose not to return to Vassar. She had ornery computers to wrestle, and she was just having too much fun.

In 1949, Hopper moved to the Eckert-Mauchly Computer Corporation in Philadelphia, where she helped design the first electronic digital computer for large-scale commercial use. She also returned to what she'd identified as a problem with programming: it was very specialized and very dull. At the time, programmers had to manually enter every 1 and 0. What the human/machine interface needed was a sort of translator, a program that would take reasonable human commands and transform them into the binary language of computers. Never the type to wait to have things done for her, Hopper designed one. Her program A-0, which stands for automatic programming language zero, is now known as the first "compiler." In the history of programming languages, adding the ability both to interact with the machine more intuitively and to pack more into a command was hugely significant. Instead of having to input strings of 1's and 0's to explain to the computer what it needed to do, Hopper condensed those strings into, say, one letter on a keyboard.

She also provided the foundation for COBOL (common business oriented language), a programming language designed specifically for business use. Even today, COBOL remains a major player in business and government organizations.

In 1966, Hopper retired from the Naval Reserve. It didn't

last long. Her presence was requested for a six-month stint to work on Automatic Data Processing, at which point the navy made it clear that her services would be required indefinitely. Hopper was promoted to captain and then, in 1977, made the special advisor to the commander of the Naval Data Automation Command. During her second stint in the navy—one that lasted nineteen years post-"retirement"—she helped set common standards for the organization's programming languages. Those standards made their way to the Department of Defense and then into all of our computers.

When Hopper, smoking unfiltered Lucky Strikes, strolled confidently down a conference hall corridor with a group trailing behind her, people routinely turned in awe. At the podium, she was a captivating visionary, exciting listeners with predictions about the future of computers and challenging the audience to think more creatively.

Once, when she was asked about the boundaries of a technology, she replied, "They'll only be limited if our imaginations are limited. It's all up to us. Remember, there were people who said the airplane couldn't fly."

INVENTION

HERTHA AYRTON

1854–1923

PHYSICS • BRITISH

WHEN EARLY THEATERGOERS NICKNAMED CINEMA "THE flicks," the name was an affectionate reference to a technological quirk. The powerful light beam directed through film strips fluttered, sending black-and-white moving images to the screen in bursts and dips. That flicker came from early projectors' arc lighting, which was created when two carbon rods placed next to each other were electrified. The electricity jumped the gap between the two rods, causing a brilliant, if unsteady, arc. Over time, arc lighting's flicker was verbally shortened to *flick*, and the name stuck despite modern cinema's steady projections.

Arc lighting dates back to 1807, but it wasn't until generators caught up with the technology's needs in the 1870s that industry could finally use it. Too bright for homes, arc lights became the go-to solution for lighthouses and other applications where very strong beams were needed. By the 1890s, they started to replace gas in streetlights, later becoming famous for their place in films, both illuminating the sets of movies like *Citizen Kane* and beaming early silent film stars to the screen.

Arc lighting should have been background, but because the lights hissed and sputtered, they claimed a prominent part in every production. The ruckus occurred in the rods. When they were electrified, the carbon evaporated and a tiny hole formed. As air rushed into the divot, it created a whine. Constantly tweaking and adjusting the rods in an effort to coax them into doing their job without too much protesting, arc light attendants were always busy.

Scientists like Hertha Ayrton, a British inventor and physicist, and her husband, William, an electrical engineer, started working toward a quieter and more consistent arc light in the late 1800s. Unfortunately, their work went up in flames when it was mistaken for kindling, crumpled by the maid, and tossed into the fireplace. (No word on whether the fire burned brighter.) The mistake occurred while her husband was away in the United States on business, so Ayrton restarted the research by herself.

She began by mounting a thorough investigation. By understanding the process's intricacies, she hoped to identify the problem and figure out how to engineer it to cut the hiss and flicker.

When she discovered that the rod was the problem, Ayrton designed one shaped for quieter use. Along the way, Ayrton also got clarity on the light's flutter, by learning about the relationship between the voltage drop across the arc, the arc's length, and the current. In 1895 and 1896 she published twelve papers in *The Electrician* that laid out her findings.

Ayrton demonstrated her work on arcs for the Royal Society in 1899. A newspaper gushed that the "lady visitors" were "astonished . . . one of their own sex [was] in charge of the most dangerous-looking of all the exhibits—a fierce arc light enclosed in glass. Mrs. Ayrton was not a bit afraid of it."

Members of the Royal Society, however, were a bit afraid of her. When Ayrton's paper "The Mechanism of the Electric Arc" was accepted in 1901, the society recruited a male member to publicly present it, as women weren't allowed. A year later, she earned a nomination to join the society, but the group consulted a lawyer who decided that her sex made her ineligible; according to English common law, a married woman had no legal standing separate from her husband.

Ayrton thought that the discrimination she faced was utter nonsense. "Personally I do not agree with sex being brought into science at all," she explained to a journalist. "The idea of 'women and science' is entirely irrelevant. Either a woman is a good scientist or she is not; in any case she should be given opportunities, and her work should be studied from the scientific, not the sex, point of view."

Ayrton was one of the good scientists. Her 450-page book, *The Electric Arc*, became the standard on arc lighting nearly as soon as it was published in 1902. But it wasn't until two years later that the Royal Society allowed Ayrton to read a paper of her own. Finally, the organization had come around. In 1906, Ayrton was awarded the society's Hughes Medal "for an original discovery in the physical sciences, particularly as applied to the generation, storage and use of energy." Membership, however, was still out of her reach.

Until 1918, women's right to vote was, too. Informed by her own early poverty and continuing experience with sexism, Ayrton was an outspoken suffragist operating with authority, charm, and presence. She cared for suffragist hunger strikers and refused to participate in the 1911 census. Scrolled across the official census form she wrote, "How can I answer all these questions if I have not the intelligence to choose between two candidates for parliament? I will not supply these particulars until I have my rights as a citizen. Votes for women. Hertha Ayrton."

Ayrton was one of a small club of women attempting to gain acceptance in the overwhelmingly male scientific institutions. Ayrton counted Marie Curie among her closest friends, and often stuck up for the chemist's reputation publicly. "An error that ascribes to a man what was actually the work of a woman has more lives than a cat," wrote Ayrton in response to a common Curie refrain. When Curie's husband, Pierre, died in 1906 and Ayrton's husband, William, died in 1908, both went on to prove that, though their husbands were valued collaborators, they possessed scientific prowess of their own.

Science was actually Ayrton's second career. Before her exploration into arc lighting, she was an inventor, patenting a device that would divide a line into equal segments. (Some biographers ascribe her affinity for tinkering to her watchmaker father.) In World War I, dismayed by reports of chlorine gas being used on British soldiers, she was drawn to invention again. The self-assigned task was this: How could she protect soldiers from the noxious gas? To experiment with a variety of methods, Ayrton staged a miniature war zone in her drawing room, with matchboxes serving as trenches and cooled smoke (produced from brown paper lit on fire) standing in as gas, which she poured over the circuit. There she refined what she believed to be the best solution—essentially a long broomstick topped by a large rectangular paddle, which would force the gas away when flapped manually.

The military was initially skeptical. What could these fans possibly do in battle? The organization's hang-ups were partly semantic. "Fans" were objects that women carried. It took a couple of years and a demonstration in the field in 1917, but the military finally put the devices to use; some one hundred thousand were eventually shipped to the Western Front. Two years later, Ayrton completed an automatic version to contend with gas carried on more powerful winds.

Ayrton was a creative problem solver. She had the flexibility and skill set to tackle a hiss, a flicker, or a deadly gas, whether it required a set of pillboxes or the principles of physics. It never mattered if others believed those things weren't within her reach. She knew they were.

HEDY LAMARR

1914–2000

TECHNOLOGY • AUSTRIAN

HEDY LAMARR KNEW WHAT WAS EXPECTED OF HER, AND BEcoming the inventor of a secret communication system—that would usher in technologies like Wi-Fi, Bluetooth, and GPS—was not it. But no one really pegged her for a Hollywood film star, either. Lamarr was, after all, born in 1914 and raised half a world away, in Vienna, Austria. Even the precocious daughter of a banker with training in dancing and piano wouldn't have a hope of landing so much success so far away. But Lamarr was never concerned about what other people believed was within her grasp or out of it. She had her own restlessness to contend with. "I've never been satisfied," said Lamarr. "I've no sooner done one thing than I am seething inside to do another thing." Even amid divorce, war, and rejection, Lamarr could spot an opening that would bring her closer to advancement, no matter how obscured.

When Lamarr (née Hedwig Kiesler) was a child, she wandered the streets of Vienna with her father, listening to him explain the inner workings of complicated machines like streetcars and printing presses. He put a high value on independence: "[My father] made me understand that I must make my own decisions, mold my own character, think my own thoughts." Not only did he provide her with marching orders to find her own way in the world; he also gave her the ammunition with which to carry them out. When Lamarr made the decision to leave school at sixteen and move to Berlin in order to pursue acting, she knew her father would not stop her.

Lamarr quickly made a name for herself on the stage and screen. But her ascent was not without snags. An early one was her marriage to a wealthy (and persistent) munitions dealer, Friedrich "Fritz" Mandl, who promptly forced her to quit her public-facing career as an actress for a new role at home: the trophy wife. Becoming an accessory used to thrill her husband's powerful friends, however, did not suit her. "Any girl can be glamorous," Lamarr said. "All you have to do is stand still and look stupid."

Before long, Lamarr began plotting her escape. While she performed her act as a well-coiffed houseplant, she paid careful attention to the sensitive conversations her husband was having with his guests, who included diplomats, politicians, generals, and Benito Mussolini. Lamarr planned to leverage what intelligence she'd gathered against her controlling husband, should he refuse to allow her to quit the marriage. It never came to that. By 1937, after Mandl stormed off to one of his hunting lodges following a fight, Lamarr left for London with two large trunks, two small ones, three suitcases, and as much jewelry as she could carry. (Money was difficult to take out of the country.) Upon arriving she was able to arrange an introduction with the head of MGM Studios, Louis B. Mayer, the executive with the largest salary in the United States. They met at a small party. Unlit cigar in hand, he chided her for a nude appearance she'd made in an art film, telling her, "I don't like what people would think about a girl who flits bare-assed around a screen." There it was again: *what people think*. He offered her a $125-a-week contract with MGM if she could find her own way to California. Lamarr turned him down. Salacious scene or not, Lamarr knew her value by the way Mayer inspected her—and it was more than he was offering.

But Lamarr also understood that Mayer was her best ticket to Hollywood, so when the MGM head and his wife hopped on a 1,028-foot ocean liner to the United States, Lamarr made sure she secured herself a spot on the ship, too. By the time the boat arrived stateside, Mayer had upped his offer: five hundred dollars a week for seven years if she agreed to English lessons and a name change. Her new moniker, decided over a Ping-Pong table while they traveled across the Atlantic Ocean, was marquee-ready. At age twenty-two, Hedwig Kiesler walked off the ship newly anointed as Hedy Lamarr. She was cast in her first Hollywood film seven months later.

As her career ramped up, Lamarr realized she wasn't especially fond of Hollywood in the off hours—too many social occasions with "people who kid all the time," she said. Lamarr preferred time to herself to tinker. Restless and still engaged in how the world worked, Lamarr transformed her drawing room into a workshop where she could fiddle with the many ideas that preoccupied her. There, she reimagined everything from tissue disposal to soda. For the latter, Lamarr convinced the high-flying manufacturing magnate Howard Hughes to loan her two chemists to help with experiments that would transform a bouillon cube into a savory cola. In *Forbes* magazine years later, Lamarr laughed about the effort: "It was a flop."

By 1940, the headlines about World War II became more serious. Just one month apart, two British ocean liners carrying children to safer waters were torpedoed by German U-boats. In the second incident, seventy-seven children were killed by people who spoke Lamarr's mother tongue. She was both shaken and incensed. She deeply wanted to find a way to help the Allied forces. Perhaps, she thought, all that information she'd gathered on German military tech might be of use in defending against the Germans.

Lamarr was so serious about getting the information to officials in her adopted country that, for a time, she considered quitting acting in order to lend her knowledge of Mandl's dealings to the National Inventors Council, a group established during World War II as a sort of clearinghouse for ideas, submitted by the public, that might help the war effort. Instead, she decided to design something practical, a technology that the military desperately needed: a better way to guide torpedoes.

By 1942, US torpedoes had a whopping 60 percent fail rate. The weapons, which were improperly tested before deployment, were tossed out like bowling balls with spin but no aim. They would often dive too deep, burst too early, or do nothing at all. On other occasions, the torpedoes hit enemy ships, but without enough oomph to sink them. The weapons needed a better in-action guide to keep them on course. Lamarr started thinking about communication. If the soldiers ordering the torpedoes could keep tabs on them en route, the effect would be like installing bumper lanes in the vast, uncertain sea. Should the missile start to veer off, a human could guide it back from afar.

Engineers had been thinking about the communication problem for decades, but they hadn't yet uncovered a solution that was enemy-proof. Although radio could offer a connection between sub and torpedo, the technology had an oversharing problem. Once a station was established, enemies could easily gum it up, jam it, or listen to the signal. The line was too public. What soldiers needed was a way to talk to their weapons without the enemy overhearing the instructions. An antijamming technique had been floated in 1898 by a US Navy engineer, but his solution—transmitting over higher and higher frequencies—wouldn't have lasted long as opposing forces one-upped each other for higher and higher real estate. Lamarr, however, had another idea about how to secure a safe and clear connection. Since setting a single frequency left the communication vulnerable, she thought that a coordinated effort where both the sender and the receiver hopped frequencies in a pattern would confound anyone trying to listen in. The idea was similar to two pianos playing in unison.

Helping her to advance the idea was Lamarr's friend George

Antheil, a composer who put together movie scores to help support his more experimental work. Antheil was famous for a piece he produced in Paris in 1926 called *Le Ballet Mécanique.* Although humans ended up playing the parts, the work called for automated player pianos to perform in sync. Lamarr, also an accomplished pianist, sometimes played recreationally with Antheil. The duo would play a game sort of like chase across the keys. One person would start playing a tune, and the other would have to catch the song and play alongside. According to her son, this synchronized musical discourse gave the inventor her idea for outsmarting the Axis opponents. Antheil, who had already put quite a lot of thought into how to synchronize machines and who had, at one point, been a US munitions inspector, was the perfect partner to help Lamarr implement her idea.

Over countless hours on the phone, in the evenings, and spread out with matchsticks and other knickknacks on Lamarr's living room rug, the pair nailed down the basics for their frequency-hopping invention. They applied for a patent in June 1941.

More concerned about the war than monetization, Lamarr and Antheil also sent their ambitious plans to Washington, DC, for review from the National Inventors Council. The positive feedback was swift. In a special to the *New York Times,* the council leaked its approval. The article began, "Hedy Lamarr, screen actress, was revealed today in a new role, that of an inventor. So vital is her discovery to national defense that government officials will not allow publication of its details." The idea was classified "red hot" by the council's engineer.

The bombing of Pearl Harbor changed the perception of the project. With the tragedy came many revelations about the sorry state of the United States' existing torpedoes. At this point, the navy decided that they had neither the bandwidth nor the interest to test another system. Lamarr and Antheil secured the patent but lost out on a government contract. Lamarr's patent was classified and filed away, its inventors' chances for real-world deployment left in the dusty back pockets of a government cabinet.

It wasn't until two decades later that the idea resurfaced, wrapped into a new frequency-hopping communication technology (later called spread-spectrum). Even then, the idea didn't go public until 1976—thirty-five years after Lamarr patented it.

As it turned out, the technology had broader uses than just missiles. Lamarr's idea paved the way for a myriad of technologies, including wireless cash registers, bar code readers, and home control systems, to name a few. While she had a long career as a celebrated actress, Lamarr finally got the full recognition she deserved when she was awarded the Electronic Frontier Foundation's Pioneer Award in 1997. Her response: "It's about time."

RUTH BENERITO

1916–2013

CHEMISTRY • AMERICAN

THE COTTON INDUSTRY WAS IN A TAILSPIN. IN 1960, IT PRO-duced a cushy 66 percent of the clothes in American homes. By 1971, cotton's market share was nearly cut in half. Nylon, polyester, and other lab-made synthetics developed in the 1930s and 1940s had charmed their way onto hangers. Sure, synthetics had drawbacks. They held on to body odor and could get itchy. But they also performed this one really outstanding trick: synthetics didn't require an iron.

Cotton's wrinkling problem was a product of the material's weak hydrogen bonds. At the molecular level, the fabric is made up of strong chains of cellulose drawn together by hydrogen. Washing the cotton caused the cellulose chains to flap around. Meanwhile, the hydrogen atoms sat idly by, doing nothing to restore the order. Even after being pulled from the line or from the dryer, cotton clothing had wrinkles. To smooth the cellulose, you needed an iron.

Morning after morning, Americans held up two shirts: one that required setting up a cloth-covered table, a hot metal object, and some spare time, and another that could be yanked from a clean laundry pile and buttoned up immediately. Synthetics were unstoppable.

Or at least it looked that way until 1969, when Ruth Benerito saved the cotton industry from collapsing. Her discovery of wrinkle-free cotton brought the material back from the brink.

It's important to note that Benerito had a habit of downplaying her abilities. On going into chemistry: "I'm not good with my hands. My mother said she didn't know why I went into chemistry 'cause I was so terrible with my hands." On discovering wrinkle-free cotton: "Any number of people worked on it."

Graceful motor skills or not, Benerito jumped into the women's college at Tulane University when she was fifteen. By the time she was nineteen in 1935, she'd earned her bachelor's in chemistry. The year was a lousy one for an aspiring chemist looking for employment. The Great Depression made it impossible for her to land a job in her field, so she took a position teaching high school and waited it out. The window of opportunity finally opened during World War II when spots vacated by men in industry and in universities were opened up to women. Benerito taught at Tulane, finally getting her PhD after the war.

Looking back on her life and education, Benerito realized she had benefited from two separate incredible moments in scientific research. The first occurred while she was attending PhD classes at the University of Chicago in the summer. "It was a good education because I was taught . . . by the greatest chemists of the last century," she mentioned nonchalantly. She was there when the university served as a Manhattan Project hub. Several of her professors were Nobel Prize winners, and some classes were so small that Benerito would be in the company of just one or two other students. "I think that's what gave me such a good background in chemistry," she said. The Cold War—"when [the government] put a lot of money into science because we were competing with Sputnik"—was also favorable for Benerito and her colleagues.

Between the two periods, she returned to Tulane to teach at the engineering school. She enjoyed watching students succeed, but eventually the promotions given to her less experienced male colleagues grated on her. When a new dean came in, she asked for a raise. He replied that he'd need some time to personally evaluate her performance. It was a blatantly obvious brush-off if she'd ever seen one. "I said I've been here thirteen years. If you don't know me now, you'll never know me," she said. "So I quit."

Some former students who'd gotten jobs at the US Department of Agriculture saw Benerito's resignation as their opening to rope a major talent. She was hired in 1953 for what would become a very productive, thirty-three-year career. The purpose of the USDA's New Orleans outpost was to push America's farm products into the future with data, science, and engineering. Benerito came to the post full of ideas and initiative.

This time, her abilities didn't go unnoticed. Within five years, Benerito was named leader of the lab that would make fabric history. Remember those breakable bonds between long cellulose chains? To strengthen those connections, Benerito experimented with shorter bonds that would "cross-link" the longer fibers, acting like a series of rungs on a ladder. When washed and dried, the cross-links would hold the long cellulose chains in place, convincing them to lie flat for wrinkle-free fibers.

She wasn't the first one to try cross-linking. But previous attempts caused cotton fibers to act strangely. Some became so rigid that just the act of sitting down could produce a Hulk-like effect, splitting the treated shirt all the way up the back.

Benerito's big innovation was in the additive. Instead of going with one that chemically attached to the cellulose chains, she found one that smoothed the surface. Her innovation not only kicked off the "wash and wear" cotton industry but also provided the foundation for stain-resistant and flame-retardant fabrics. Benerito earned the Lemelson-MIT Lifetime Achievement Award and the USDA's highest honor for service—twice!

Though she would feel uncomfortable claiming the title, the Queen of Cotton had been crowned.

STEPHANIE KWOLEK

1923–2014

CHEMISTRY • AMERICAN

IN A PAPER PUBLISHED IN 1959, TITLED "THE ROPE TRICK," STEPHanie Kwolek and a coauthor explain how, with the right chemicals, anyone can produce the chemistry equivalent of "pulling a string of silk handkerchiefs out of a top hat." To magically pull nylon from a beaker, first you layer diacid chloride and a solvent on top of an equal amount of diluted aliphatic diamine, which sit together like oil and water. But dip a wand into the intersection of the two fluids and pull up—and voilà! A net of nylon appears like a circus tent, gathering at the top to form a string. So much of the stuff can be lifted from the solution that one modern experimenter attached the thread to an automatic drill and let it coil continuously around the bit.

The reaction was an impressive piece of chemical showmanship, but Kwolek's next trick would be death-defying. In 1964, she designed a fabric that could stop a speeding bullet.

If someone had asked Kwolek as a child what she believed she'd be doing as an adult, it was not this. When she was young, Kwolek loved fabrics and sewing, so she imagined that one day she might be a fashion designer. Kwolek's mother talked her out of it, fearing that Kwolek's tendency toward perfection would lead her to starve should she find herself unsatisfied with a hem. After cultivating a love for science, Kwolek changed her mind, hoping for a career in medicine.

In 1946, she graduated with a degree in chemistry from Carnegie Mellon University in Pittsburgh. Without any luck with loans, Kwolek had to put medical school aside until she could afford it. Thankfully, Kwolek was hired by DuPont as a chemist right out of college. After going in for an interview, she asked her would-be boss to speed up his decision, as she had another opportunity pending. He prepared the offer letter then and there. She later mused that it was her assertive style that got her the position.

The idea was that Kwolek would put in a few years at DuPont in order to save the money she needed to become a doctor. But a funny thing happened on the way to medical school. Kwolek may not have been designing clothes, but she was using chemicals to create new, futuristic fabrics. The threads she set out to create would challenge the concept of what a material is capable of—a feat that would change the course of history. When Kwolek compared her opportunities at DuPont to what she'd get out of medical school, her earlier aspirations faded. Kwolek's adventures in chemistry were just too rewarding to give up.

Furthermore, DuPont was in the middle of a particularly vibrant period. The company was experimenting with all sorts of ways to make synthetic materials mimic the incredible properties of nature. Spider silk's strength and elasticity, for instance, was the inspiration for the invention of nylon in the 1930s. Even thirty years later, the company continued to strive for better synthetics. In the 1960s, DuPont tasked Kwolek with designing a replacement for steel reinforcement in tires. They needed a material that was both lighter and stronger.

In this task, Kwolek attempted to make a liquid polymer by combining two crystallized ones. Typically when polymer A was mixed with polymer B, it created a clear, thick goop, which could be spun into string. But when Kwolek repeated the process at lower temperatures, the result was a liquid—neither goopy nor clear. Repeating the experiment under the same conditions, Kwolek got an identical result. Her colleagues were dubious. The foggy mixture looked like a candidate for the trash, not one for production. The technician responsible for spinning liquid polymers into string initially hemmed and hawed, fearing that the liquid might gunk up his machine. But Kwolek stood by her work and pushed to take it further. The result was a thread of incredible lightness and strength never before seen in a lab. In 1964, Kwolek invented Kevlar.

"It wasn't exactly a 'eureka moment,'" Kwolek admitted to a local paper. Even though the fiber's readings were off the charts—five times as strong as steel and most certainly lighter—she wanted to be absolutely sure that she had her data in order because once she revealed her results to the company, she had a pretty good feeling that DuPont would immediately throw resources at the project.

Even after showing her results, Kwolek admitted, "I never in a thousand years expected that little liquid crystal to develop into what it did." With the attention of an entire team at DuPont, Kevlar's properties became even more remarkable.

Because of its strength and exceptional light weight, Kevlar has been applied to everything from oven gloves to space suits to cellphones. In bulletproof vests, Kevlar has protected some three thousand law enforcement officers from bullets.

Kwolek's preparation of the cold-spun threads launched a brand-new area of research around liquid crystalline polymers. For her work on Kevlar and her subsequent contributions to Lycra and Spandex, Kwolek won the Lemelson-MIT Lifetime Achievement Award in 1999.

Whipping that strange, cloudy liquid into super-strong strings proved to be an extraordinary trick.

ACKNOWLEDGMENTS

MANY OF THE EXTRAORDINARY STORIES OF DISCOVERY, CREativity, bravery, and grit would not have been featured here were it not for scholars like Marilyn B. Ogilvie and writers like Sharon Bertsch McGrayne keeping them in the public discourse.

Domenica Alioto offered sage advice, calming support, and brilliant feedback to the project. Mackenzie Brady's encouragement and expert ability to connect the dots made *Headstrong* possible. Matt Weiland believed in Yvonne Brill and the gang from the beginning.

Tim Leong, Sharon Swaby, Gordon Lindsay, Elise Craig, Bryan Lufkin, Jordan Crucchiola, Lydia Belanger, Lexi Pandell, Julia Greenberg, Kevin Newcomen, Laurie Prendergast, and Brian Moyers offered invaluable feedback before publication. Stephen Swaby, Sean Swaby, Holly Brickley, Dan Lyon, John Talaga, Emily Rolph, Amy Cooper, Caitlin Roper, Shirley Lindsay, Nancy Leong, Rich Leong, and Courtney Hughes gave much-needed encouragement and support.

Thank you to Yvonne Brill for making a mean beef stroganoff . . . and for keeping communications satellites from slipping out of orbit. While discussion of the former helped get this book off the ground, here's hoping she'll be remembered for the latter.

NOTES

INTRODUCTION

xi *"world's best mom"* Douglas Martin, "Yvonne Brill, a Pioneering Rocket Scientist, Dies at 88." *New York Times,* March 30, 2013.

xi *"What astonished the lady visitors"* Evelyn Sharp, *Hertha Ayrton: 1854–1923, a Memoir.* London: E. Arnold & Company, 1926.

xii *"The idea of 'woman and science'"* Ibid.

xii *"As a parent"* As quoted in Lynn Sherr, *Sally Ride: America's First Woman in Space.* New York: Simon & Schuster, 2014.

xii *"I went to a store"* Charlotte to Lego Company, January 25, 2014, in *Sociological Images,* http://thesocietypages.org/socimages/2014/01/31/this-month-in-socimages-january-2014/.

MARY PUTNAM JACOBI

3 *"There have been instances"* Edward H. Clarke, *Sex in Education or, A Fair Chance for Girls.* Boston: James R. Osgood and Company, 1875.

3 *"crowds of pale"* Ibid.

4 *"I really am only enjoying"* Mary Putnam Jacobi, *Life and Letters of Mary Putnam Jacobi.* New York: G. P. Putnam's Sons, 1925.

4 *"There is nothing"* Mary Putnam Jacobi, *The Question of Rest for Women During Menstruation.* New York: G. P. Putnam's Sons, 1877.

5 *"I began my medical studies"* Mary Putnam Jacobi, *Life and Letters of Mary Putnam Jacobi.* New York: G. P. Putnam's Sons, 1925.

5 *"I think you are rather"* Ibid.

6 *"I . . . feel as much at home"* Ibid.

ANNA WESSELS WILLIAMS

7 *"near-epidemic levels"* John Emrich, "Anna Wessels Williams, M.D.: Infectious Disease Pioneer and Public Health Advocate." *AAI Newsletter,* March/April 2012.

8 *"happy to have the honor"* "Dr. Anna Wessels Williams." *Changing the Face of Medicine.* National Library of Medicine, www.nlm.nih.gov/changingthefaceofmedicine/physicians/biography_331.html. accessed November 1, 2013.

9 *"I was starting"* The quote appeared in Regina Markell Morantz-Sanchez, *Sympathy & Science: Women Physicians in American Medicine.* Chapel Hill: University of North Carolina Press, 2000.

10 *"to find out about"* Ibid.

10 *"a scientist of international repute"* "94 Retired by City; 208 More Will Go." *New York Times,* March 24, 1934.

ALICE BALL

11 *"The pit of Hell"* Jack London, *The Cruise of the Snark.* New York: Macmillan, 1911.

GERTY RADNITZ CORI

14 *"As a research worker"* Gerty Cori, *This I Believe,* hosted by Edward R. Murrow, September 2, 1952.

17 *"The lab made so many discoveries"* Sharon Bertsch McGrayne, *Nobel Prize Women in Science: Their Lives, Struggles, and Momentous Discoveries.* 2nd ed. Washington, DC: National Academies Press, 2001.

18 *"for their discovery"* "Gerty Cori — Facts," Nobel Prize, http://www.nobelprize.org/nobel_prizes/medicine/laureates/1947/cori-gt-facts.html, accessed August 20, 2014.

HELEN TAUSSIG

19 *"the crossword puzzle"* As quoted in Jody Bart, *Women Succeeding in the Sciences: Theories and Practices Across Disciplines.* West Lafayette, IN: Purdue Research Foundation, 2000.

20 *"Who is going to be such a fool"* Ibid.

21 *"I close ductuses"* Ibid.

21 *"I suppose nothing would ever give"* Ibid.

21 *"Dr. Gross [of Harvard]"* Ibid.

22 *"the knights of Taussig"* Ibid.

22 *"You have your sadnesses"* Jeanne Hackley Stevenson, "Helen Brooke Taussig, 1898: The 'Blue Baby' Doctor." *Notable Maryland Women.* Cambridge, MD: Tidewater, 1977.

ELSIE WIDDOWSON

25 *"The modern loaf"* Jane Elliott, "Elsie—Mother of the Modern Loaf." *BBC News,* March 25, 2007.

26 *"Tender loving care"* As quoted in Margaret Ashwell, "Elsie May Widdowson, C.H., 21 October 1906–14 June 2000." *Biographical Memoirs of Fellows of the Royal Society,* December 1, 2002.

26 *"A slight accident"* Ibid.

VIRGINIA APGAR

27 *"Nobody, but nobody"* "The Virginia Apgar Papers: Biographical Information." US National Library of Medicine. http://profiles.nlm.nih.gov/ps/retrieve/Narrative/CP/p-nid/178, accessed June 13, 2014.

27 *"never sat down"* Ibid.

27 *"Frankly, how does she do it"* Ibid.

29 *"That's easy"* "The Virginia Apgar Papers: Obstetric Anesthesia and a Scorecard for Newborns, 1949–1958." US National Library of Medicine. http://profiles.nlm.nih.gov/ps/retrieve/Narrative/CP/p-nid/178, accessed June 13, 2014.

30 *"people doctor"* "The Virginia Apgar Papers: The National Foundation–March of Dimes, 1959–1974." US National Library of Medicine. http://profiles.nlm.nih.gov/ps/retrieve/Narrative/CP/p-nid/178, accessed June 13, 2014.

31 *"Her warmth and interest"* Joseph F. Nee, memorial service, "The Virginia Apgar Papers." US National Library of Medicine. New York, September 15, 1974.

DOROTHY CROWFOOT HODGKIN

33 *"gentle genius"* As quoted in Sharon Bertsch McGrayne, *Nobel Prize Women in Science: Their Lives, Struggles, and Momentous Discoveries.* 2nd ed. Washington, DC: National Academies Press, 2001.

34 *"just the right size"* Ibid.

35 *"broke the sound barrier"* Ibid.

35 *"For her determinations"* "The Nobel Prize in Chemistry 1964." Nobel Prize. http://www.nobelprize.org/nobel_prizes/chemistry/laureates/1964/, accessed August 15, 2014.

GERTRUDE BELLE ELION

36 *"was the turning point"* As quoted in Sharon Bertsch McGrayne, *Nobel Prize Women in Science: Their Lives, Struggles, and Momentous Discoveries.* 2nd ed. Washington, DC: National Academies Press, 2001.

36 *"distracting influence"* Ibid.

37 *"I've learned whatever you have"* Ibid.

38 *"Oh, no, I'm not quitting that job"* Ibid.

39 *"This is the best thing"* Ibid.

40 *"In fifty years"* Ibid.

JANE WRIGHT

41 *"Lollygagging"* Alison Jones, personal interview, September 14, 2014.

41 *"the mother of chemotherapy"* Ronald Piana, "Jane Cooke Wright, MD, ASCO Cofounder, Dies at 93." *ASCO Post*, March 15, 2013.

41 *"renowned artist"* "Homecoming for Jane Wright." *Ebony*, May 1968.

42 *"His being so good"* As quoted in Lisa Yount, *Black Scientists*. New York: Facts on File, 1991.

42 *"the Cinderella"* Jane C. Wright, "Cancer Chemotherapy: Past, Present, and Future—Part I." *JAMA*, August 1984.

42 *"It is almost, not quite, but almost"* Ibid.

44 *"She was one of the few people"* Alison Jones, personal interview, September 14, 2014.

JEANNE VILLEPREUX-POWER

52 *"power cages"* Jeannette Power, "Observations on the Habits of Various Marine Animals." In *Annals and Magazine of Natural History*. London: Taylor and Francis, 1857.

53 *"one of the most eminent"* Matilda Joslyn Gage, "Woman as an Inventor." In *The North American Review*, edited by Allen Thorndike Rice. New York: AMS Press, 1883.

MARY ANNING

54 *"lively and intelligent"* Charles Dickens, "Mary Anning, The Fossil Finder." *All The Year Round: A Weekly Journal*, July 22, 1865.

56 *"I beg your pardon"* Ibid.

56 *"The carpenter's daughter"* Ibid.

ELLEN SWALLOW RICHARDS

58 *"that her admission did not"* Records of the Meetings of the MIT Corporation, December 14, 1870, *Archival Collection AC*.

58 *"Had I realized"* Caroline Louisa Hunt, *The Life of Ellen H. Richards*. Boston: Whitcomb & Barrows, 1912.

58 *"I have felt the greatest"* First Annual Report to the Women's Educational Association circa 1877, folder 9, Collection on the Massachusetts Institute of Technology Women's Laboratory, 1867–1922 (AC 0298), Institute Archives and Special Collections, MIT Libraries, Cambridge, Massachusetts.

ALICE HAMILTON

61 *"fourth-rate bacteriologist"* Alice Hamilton, *Exploring the Dangerous Trades.* Boston: Little, Brown, 1943.

61 *"happy dust"* Ibid.

61 *"as if I was going up in a flying machine"* Ibid.

62 *"So I tested the powders on myself"* Ibid.

62 *"As I prowled about the streets"* Ibid.

63 *"For years"* Ibid.

64 *"poisonous occupations"* Ibid.

64 *"tin foil"* Ibid.

64 *"leaded"* Ibid.

65 *"I was about the only"* Ibid.

65 *"No young doctor nowadays"* Ibid.

ALICE EVANS

67 *"was almost universal skepticism"* Alice C. Evans, *Memoirs.* Unpublished. 1963. In Alice Catherine Evans, Papers #2552, Division of Rare and Manuscript Collections, Cornell University Library.

TILLY EDINGER

70 *"like an ammonite in the Holocene"* Emily A. Buchholtz and Ernst-August Seyfarth, "The Gospel of the Fossil Brain: Tilly Edinger and the Science of Paleoneurology." *Brain Research Bulletin*, 1999.

70 *"fighting like a hero"* Ibid.

71 *"The fossil vertebrates will save me"* Ibid.

71 *"She is a research scientist of the first rank"* Ibid.

73 *"The Tilly Bat"* Glenn Jepsen, letter to Tilly Edinger, June 11, 1957.

RACHEL CARSON

76 *"I had given up writing forever"* Rachel Carson. *Lost Woods: The Discovered Writing of Rachel Carson*, edited by Linda Lear. Boston: Beacon Press, 1998.

77 *"Every living thing of the ocean"* Ibid.

77 *"We live in a scientific age"* Ibid.

78 *"Almost immediately DDT was hailed"* Rachel Carson, *Silent Spring.* New York: Houghton Mifflin, 1962.

78 *"efforts which will prevent"* "The National Environmental Policy Act of 1969," 42 U.S.C., January 1, 1970.

78 *"the most important piece"* Jack Lewis. "The Birth of EPA." *EPA Journal*, November 1985.

RUTH PATRICK

80 *"Come here at once"* "Almost Had a War on Her Hands." *Sydney Morning Herald*, August 11, 1960.

80 *"So you're the lady"* Ibid.

80 *"You see, diatoms are like detectives"* Sandy Bauers, "Ruth Patrick: 'Den Mother of Ecology.'" *Philadelphia Inquirer*, March 5, 2007.

81 *"collected everything"* Ibid.

82 *"a little peon"* Ibid.

82 *"You can't have society without industry"* Michael Roddy, "Pollution Fears Come to Lakes, Springs," Associated Press, January 8, 1984.

82 *"I try not to think about it"* Sandy Bauers, "Ruth Patrick: 'Den Mother of Ecology.'" *Philadelphia Inquirer*, March 5, 2007.

NETTIE STEVENS

87 *"I beg to urge"* Stephen G. Brush, "Nettie M. Stevens and the Discovery of Sex Determination by Chromosomes." *Isis*, June 1978.

87 *"Here . . . it is perfectly clear"* Bryn Mawr College Monographs, Reprint Series, vol. 7, N. M. Stevens, *Studies in Spermatogenesis, Part I & Part II*, Washington DC: Carnegie Institution of Washington, 1905 & 1906.

HILDE MANGOLD

89 *"The large, especially technical, difficulties"* Viktor Hamburger, "Hilde Mangold: Co-Discoverer of the Organizer." *Journal of the History of Biology*, Spring 1984.

89 *"Very few of her contemporaries"* Ibid.

89 *"We cared more about food"* Ibid.

90 *"initiated a new epoch"* Klaus Sander and Peter E. Faessler, "Introducing the Spemann-Mangold Organizer: Experiments and Insights That Generated a Key Concept in Developmental Biology." *International Journal of Developmental Biology*, 2001.

CHARLOTTE AUERBACH

91 *"We used up a lot of technicians"* As quoted in G. H. Beale, "Charlotte Auerbach, 14 May 1899–17 March 1994." *Biographical Memoirs of Fellows of the Royal Society*, November 1995.

92 *"We are thrilled by your major discovery"* Ibid.

92 *"You are the mother of mutagenesis"* Ibid.

93 *"one of the few great spiritual experiences"* Ibid.

93 *"No, I'm sorry, I'm no good at cytology"* Ibid.

BARBARA MCCLINTOCK

96 *"I didn't belong to that family"* As quoted in Sharon Bertsch McGrayne, *Nobel Prize Women in Science: Their Lives, Struggles, and Momentous Discoveries*. 2nd ed. Washington, DC: National Academies Press, 2001.

96 *"I had it done within two"* As quoted in Evelyn Fox Keller, *A Feeling for the Organism: The Life and Work of Barbara McClintock*. New York: Henry Holt, 1983.

97 *"Hell, it was so damn obvious"* Ibid.

97 *"lapping up the stimulation she provided"* As quoted in Sharon Bertsch McGrayne, *Nobel Prize Women in Science: Their Lives, Struggles, and Momentous Discoveries*. 2nd ed. Washington, DC: National Academies Press, 2001.

97 *"very powerful work"* Ibid.

98 *"You're not conscious of anything else"* As quoted in Evelyn Fox Keller, *A Feeling for the Organism: The Life and Work of Barbara McClintock*. New York: Henry Holt, 1983.

98 *"like a lead balloon"* As quoted in Sharon Bertsch McGrayne, *Nobel Prize Women in Science: Their Lives, Struggles, and Momentous Discoveries*. 2nd ed. Washington, DC: National Academies Press, 2001.

99 *"I was startled"* Ibid.

99 *"All the surprises"* Press conference on 1983 Nobel Prize, Ibid.

99 *"discovery of mobile genetic elements"* "The Nobel Prize in Physiology or Medicine 1983." Nobel Prize, http://www.nobelprize.org/nobel_prizes/medicine/laureates/1983/, accessed August 4, 2014.

SALOME GLUECKSOHN WAELSCH

100 *"further and wider"* Davor Solter, "In Memoriam: Salome Gluecksohn-Waelsch (1907–2007)." *Developmental Cell*, January 2008.

100 *"I am convinced"* Salome Waelsch, "The Causal Analysis of Development in the Past Half Century: A Personal History." *Development*, 1992.

100 *"help[ed] me more"* Susan A. Ambrose et al., *Journeys of Women in Science and Engineering: No Universal Constants*. Philadelphia: Temple University Press, 1997.

101 *"Our first meeting"* Salome Waelsch, "The Causal Analysis of Development in the Past Half Century: A Personal History." *Development*, 1992.

102 *"carries out an 'experiment'"* S. Gluecksohn-Schoenheimer, "The Development of Two Tailless Mutants." *Genetics*, 1938.

103 *"I personally am increasingly impressed"* Salome Waelsch, "The Causal Analysis of Development in the Past Half Century: A Personal History." *Development*, 1992.

RITA LEVI-MONTALCINI

106 *"I have no particular intelligence"* As quoted in Sharon Bertsch McGrayne, *Nobel Prize Women in Science: Their Lives, Struggles, and Momentous Discoveries*. 2nd ed. Washington, DC: National Academies Press, 2001.

107 *"The moment you stop"* Ibid.

ROSALIND FRANKLIN

108 *"might have been"* James Watson, *The Double Helix: A Personal Account of the Discovery of the Structure of DNA*. New York: Scribner, 1968.

108 *"a mean, mean book"* As quoted in Sharon Bertsch McGrayne, *Nobel Prize Women in Science: Their Lives, Struggles, and Momentous Discoveries*. 2nd ed. Washington, DC: National Academies Press, 2001.

108 *"unbelievably mean in spirit"* Ibid.

109 *"alarmingly clever"* As quoted in Brenda Maddox, *Rosalind Franklin: The Dark Lady of DNA*. New York: HarperCollins, 2002.

109 *"You frequently state"* Ibid.

110 *"solid organic colloids"* Ibid.

ANNE MCLAREN

114 *"everything involved"* Azim Surani and Jim Smith, "Anne McLaren (1927–2007)." *Nature*, August 16, 2007.

115 *"Four bottled babies"* John D. Biggers, "Research in the Canine Block." *International Journal of Developmental Biology*, 2001.

115 *"impeccable clarity"* Brigid Hogan, "From Embryo to Ethics: A Career in Science and Social Responsibility: An Interview with Anne McLaren." *International Journal of Developmental Biology*, 2001.

LYNN MARGULIS

116 *"I saw in her the same"* Lynn Margulis, "Chapter 7: Lynn Margulis." In John Brockman [ed.], *The Third Culture: Beyond the Scientific Revolution*. New York: Simon & Schuster, 1995.

116 *"Science was a way"* Rutgers Research Channel, "Lynn Margulis 2004 Rutgers Interview." https://www.youtube.com/watch?v=b8xqu_TlQPU>, accessed January 2013.

117 *"It just felt right"* Ibid.

117 *"Classes were not required"* Ibid.

118 *"she was scoffed at"* Lynn Margulis, "Chapter 7: Lynn Margulis." In John Brockman [ed.], *The Third Culture: Beyond the Scientific Revolution*. New York: Simon & Schuster, 1995.

118 *"It's delicious"* Ibid.

118 *"one of the classics"* Ibid.

118 *"I don't consider my ideas"* Dick Teresi, "Discover Interview: Lynn Margulis Says She's Not Controversial, She's Right." *Discover*, June 17, 2011.

ÉMILIE DU CHÂTELET

121 *"phenomenon"* Judith P. Zinsser, *Émilie du Châtelet: Daring Genius of the Enlightenment*. New York: Penguin Books, 2007.

121 *"a genius worthy of Horace"* Ibid.

122 *"sometimes happens that work"* Émilie du Châtelet, "Translator's Preface for *The Fable of the Bees*." In Judith P. Zinsser (ed.) *Émilie du Châtelet: Selected Philosophical and Scientific Writings*. Chicago: University of Chicago Press, 2009.

122 *"by a young Lady of high rank"* Judith P. Zinsser, *Émilie du Châtelet: Daring Genius of the Enlightenment*. New York: Penguin Books, 2007.

LISE MEITNER

125 *"Our Madame Curie"* As quoted in Sharon Bertsch McGrayne, *Nobel Prize Women in Science: Their Lives, Struggles, and Momentous Discoveries*. 2nd ed. Washington, DC: National Academies Press, 2001.

125 *"where they appear"* Ibid.

126 *"Nothing else has to be done"* Ibid.

128 *"Perhaps you can propose"* Ibid.

129 *"Life need not be easy"* Lise Meitner, "Looking Back." *Bulletin of the Atomic Scientists*, November 1964.

IRÈNE JOLIOT-CURIE

130 *"In our family"* Interview with Lew Kowarski by Charles Weiner, Niels Bohr Library & Archives, American Institute of Physics, College Park, MD, www.aip.org/history/ohilist/4717_1.html.

132 *"Nearly a thousand people"* "Mlle. Curie Reads Thesis." *New York Times*, March 31, 1925.

132 *"I consider science"* As quoted in Sharon Bertsch McGrayne, *Nobel Prize Women in Science: Their Lives, Struggles, and Momentous Discoveries*. 2nd ed. Washington, DC: National Academies Press, 2001.

132 *"I discovered in this girl"* Ibid.

134 *"I am not afraid of death"* Ibid.

MARIA GOEPPERT MAYER

135 *"probably the most brilliant gathering"* Max Born, *My Life: Recollections of a Nobel Laureate*. New York: Scribner, 1978.

137 *"But I don't know anything"* Robert G. Sachs, "Maria Goeppert Mayer." In Edward Shils [ed.], *Remembering the University of Chicago: Teachers, Scientists, and Scholars*. Chicago: University of Chicago Press, 1991.

137 *"Is there any evidence"* Maria Goeppert Mayer, "The Shell Model." Nobel Lecture, December 12, 1963.

138 *"Anyone who has danced"* Sharon Bertsch McGrayne, *Nobel Prize Women in Science: Their Lives, Struggles, and Momentous Discoveries*. 2nd ed. Washington, DC: National Academies Press, 2001.

138 *"if you love science"* Ibid.

MARGUERITE PEREY

139 *"In those days"* "Madame Curie's Assistant: Scientific Battle Won, She's Losing Medical One." *Milwaukee Journal*, July 15, 1962.

139 *"Under Marie Curie"* Ibid.

142 *"You are the second"* Ibid.

CHIEN-SHIUNG WU

143 *"I'm sorry, but if you wanted"* As quoted in Sharon Bertsch McGrayne, *Nobel Prize Women in Science: Their Lives, Struggles, and Momentous Discoveries*. 2nd ed. Washington, DC: National Academies Press, 2001.

144 *"you must know the purpose"* Ibid.

144 *"Equipment all alone"* Ibid.

145 *"is as obsessed with physics"* Ibid.

146 *"These are moments"* Ibid.

146 *"This small modest woman"* Ibid.

ROSALYN SUSSMAN YALOW

147 *"Personally, I have not been terribly bothered by it"* Sharon Bertsch McGrayne, *Nobel Prize Women in Science: Their Lives, Struggles, and Momentous Discoveries.* 2nd ed. Washington, DC: National Academies Press, 2001.

148 *"eerie extrasensory perception"* Ibid.

151 *"big deal"* Ibid.

MARIA MITCHELL

155 *"One gets attached"* Maria Mitchell, *Maria Mitchell: Life, Letters, and Journals.* Boston: Lee & Shepard, 1896.

155 *"sweeping the heavens"* Ibid.

156 *"For a few days"* Ibid.

156 *"It is really amusing"* Ibid.

156 *"I asked him"* Ibid.

156 *"It meant so much"* Ibid.

ANNIE JUMP CANNON

158 *"Father was more interested"* "Delaware Daughter Star Gazer." *Delmarva Star*, March 11, 1934.

158 *"smoking like a small engine"* Ibid.

160 *"My success"* Ibid.

INGE LEHMANN

161 *"the only Danish seismologist"* Edmond A. Mathez, *Earth: Inside and Out.* New York: New Press, 2000.

162 *"He gave me a great deal"* Nina Byers and Gary Williams, *Out of the Shadows: Contributions of Twentieth-Century Women to Physics.* New York: Cambridge University Press, 2006.

162 *"Never seen a seismograph"* Bruce A. Bolt, "Inge Lehmann," *Contributions of Women to Physics.* http://www.physics.ucla.edu/~cwp/articles/bolt.html, accessed September 11, 2014.

163 *"Well, the way it works"* Interview of Jack Oliver by Ron Doel on September 27, 1997. Niels Bohr Library & Archives, American In-

stitute of Physics, College Park, MD, http://www.aip.org/history/ohilist/6928_2.html.

163 *"No difference between"* S. G. Brush, "Discovery of the Earth's Core." *Am. J. Phys.* 1980. As cited in Bruce A. Bolt, "Inge Lehmann," *Contributions of Women to Physics.* http://www.physics.ucla.edu/~cwp/articles/bolt.html, accessed September 11, 2014.

164 *"Every prominent seismologist"* Interview of Jack Oliver by Ron Doel on September 27, 1997. Niels Bohr Library & Archives, American Institute of Physics, College Park, MD, http://www.aip.org/history/ohilist/6928_2.html.

MARIE THARP

165 *"great sea-gash"* "Ocean Explorer: Soundings, Sea-Bottom, and Geophysics." National Oceanic and Atmospheric Administration. http://oceanexplorer.noaa.gov/history/quotes/soundings/soundings.html, accessed September 10, 2014.

165 *"a form of scientific heresy"* Marie Tharp. "Connect the Dots: Mapping the Seafloor and Discovering the Mid-Ocean Ridge." In *Lamont-Doherty Earth Observatory of Columbia: Twelve Perspectives on the First Fifty Years 1949–1999*, edited by Laurence Lippsett. Palisades, NY: Lamont-Doherty Earth Observatory of Columbia University, 1999.

165 *"girl talk"* Ibid.

166 *"map-making in my blood"* Ibid.

166 *"bored as hell"* Hali Felt, *Soundings: The Story of the Remarkable Woman Who Mapped the Ocean Floor.* New York: Henry Holt, 2012.

166 *"a once-in-the-history-of-the-world opportunity"* Ibid.

166 *"No echo returned"* Ibid.

167 *"black cliffs in blue water"* Ibid.

168 *"Establishing the rift valley"* Ibid.

YVONNE BRILL

169 *"I didn't really discuss it"* Deborah Rice, "Interview with Yvonne Brill on November 3, 2005." Society of Women Engineers. http://www.djgcreate.com/swe/joomla/images/stories/brill/BRILLBRILL.pdf, accessed October 26, 2013.

170 *"There was just no question"* Ibid.

170 *"the cat's meow"* Ibid.

170 *"looking at the performance"* Ibid.

171 *"I never was afraid"* Ibid.

171 *"She truly represented"* American Institute of Aeronautics and Astronautics. "AIAA Mourns the Death of Honorary Fellow Yvonne C. Brill." https://www.aiaa.org/SecondaryTwoColumn.aspx?id=16827, accessed December 11, 2013.

172 *"We believe in quality"* Deborah Rice, "Interview with Yvonne Brill on November 3, 2005." Society of Women Engineers. http://www.djgcreate.com/swe/joomla/images/stories/brill/BRILLBRILL.pdf, accessed October 26, 2013.

SALLY RIDE

174 *"one thing I probably share"* Cody Knipfer, "Sally Ride and Valentina Tereshkova: Changing the Course of Human Space Exploration." NASA. http://www.nasa.gov/topics/history/features/ride_anniversary.html#.VDwXddR4pfF, accessed August, 30, 2014.

174 *"Weightlessness is a great equalizer"* "An Interview with Sally Ride." *Nova* PBS. https://www.youtube.com/watch?v=yb6vw9AmiLs, accessed August 30, 2014.

174 *"a girl physics major"* Lynn Sherr, *Sally Ride: America's First Woman in Space*. New York: Simon & Schuster, 2014.

174 *"unconscious (I assume) bias"* Ibid.

176 *"directly addresses the problems"* Sally Ride, *NASA: Leadership and America's Future in Space*, August 1987.

176 *"a better weather report"* Lynn Sherr, *Sally Ride: America's First Woman in Space*. New York: Simon & Schuster, 2014.

ADA LOVELACE

184 *"the engine might compose"* L. F. Menabrea, "Sketch of the Analytical Engine Invented by Charles Babbage, Esq.," trans. Augusta Ada Byron King, Countess of Lovelace, *Scientific Memoirs*, 1843.

184 *"It can follow analysis"* Ibid.

184 *"the Analytical Engine weaves"* Ibid.

184 *"the most important paper"* Allen G. Bromley. "Introduction." In H. P. Babbage, Volume 2: Babbage's Calculating Engines. Cambridge, MA: MIT Press, 1984. As cited in Ronald K. Smeltzer, Robert J. Ruben, and Paulette Rose, *Extraordinary Women in Science & Medicine: Four Centuries of Achievement*. New York: Grolier Club, 2013.

185 *"[not] the first woman"* Suw Charman-Anderson, "Ada Lovelace: Victorian Computing Visionary." Finding Ada. http://findingada.com/book/ada-lovelace-victorian-computing-visionary/, accessed August 29, 2014.

185 *"I am much annoyed"* As quoted in Dorothy Stein, *Ada: A Life and Legacy.* Cambridge, MA: MIT Press, 1985.

185 *"That brain of mine"* Ibid.

185 *"all this was impossible"* Ibid.

FLORENCE NIGHTINGALE

187 *"the symptoms or the sufferings"* Florence Nightingale, *Collected Works of Florence Nightingale.* Waterloo, Ontario, Canada: Wilfrid Laurier University Press, 2009.

SOPHIE KOWALEVSKI

189 *"I would stand by the wall"* Sofya Kovalevskaya, *A Russian Childhood*, trans. Beatrice Stillman, assisted by P. Y. Kochina. New York: Springer, 1978.

190 *"learned women"* Ibid.

190 *"I was in a chronic state"* Ibid.

190 *"The meaning of these concepts"* Ibid.

191 *"Sofya immediately attracted"* Ibid.

193 *"The value"* Ibid.

193 *"brain of the deceased"* Ibid.

EMMY NOETHER

194 *"Frl. Noether is continually"* As quoted in Sharon Bertsch McGrayne, *Nobel Prize Women in Science: Their Lives, Struggles, and Momentous Discoveries.* 2nd ed. Washington, DC: National Academies Press, 2001.

195 *"She won't be allowed to become a lecturer"* Ibid.

195 *"On receiving the new work"* Ibid.

195 *"an extraordinary professor"* Ibid.

195 *"You can make a strong case that her theorem"* Natalie Angier, "The Mighty Mathematician You've Never Heard Of." *New York Times*, March 26, 2012.

196 *"is as if she were describing"* Sharon Bertsch McGrayne, *Nobel Prize Women in Science: Their Lives, Struggles, and Momentous Discoveries.* 2nd ed. Washington, DC: National Academies Press, 2001.

197 *"Her heart knew no malice"* Ibid.

197 *"Fräulein Noether was the most significant"* Albert Einstein, "The Late Emmy Noether." *New York Times*, May 4, 1935.

MARY CARTWRIGHT

199 *"I heard you were extolling"* As quoted in Freeman J. Dyson, "Mary Lucy Cartwright." In Nina Byers and Gary Williams [eds.], *Out of the Shadows: Contributions of Twentieth-Century Women to Physics*. New York: Cambridge University Press, 2006.

199 *"help in solving certain very objectionable"* Donald J. Albers and Gerald L. Alexanderson, *Fascinating Mathematical People: Interviews and Memoirs*. Princeton, NJ: Princeton University Press, 2011. p. 142.

201 *"I saw the beauty"* Freeman J. Dyson, "Mary Lucy Cartwright." In Nina Byers and Gary Williams [ed.], *Out of the Shadows: Contributions of Twentieth-Century Women to Physics*. New York: Cambridge University Press, 2006.

202 *"No. Twice will do"* Philip J. Davis, "Snapshots of a Lively Character: Mary Lucy Cartwright, 1900–1998." *Society for Industrial and Applied Mathematics*. http://www.siam.org/news/news.php?id=863, accessed September 12, 2014.

202 *"Mathematics is a young person's game"* Donald J. Albers and Gerald L. Alexanderson, *Fascinating Mathematical People: Interviews and Memoirs*. Princeton, NJ: Princeton University Press, 2011.

GRACE MURRAY HOPPER

203 *"the most damaging phrase"* Diane Hamblen, Grace M. Hopper, and Elizabeth Dickason, "Biographies in Naval History: Rear Admiral Grace Murray Hopper, USN, 9 December 1906–1 January 1992." Naval History and Heritage Command. http://www.history.navy.mil/bios/hopper_grace.htm, accessed August 20, 2014.

203 *"come back and haunt"* Ibid.

203 *"It's always easier to ask forgiveness"* Ibid.

204 *"Where have you been"* Uta C. Merzbach, "Computer Oral History Collection, Grace Murray Hopper (1906–1992)." Computer Oral History Collection, 1969–1973, 1977, Archives Center, National Museum of American History, July 1968.

206 *"They'll only be limited"* Diane Hamblen, Grace M. Hopper, and Elizabeth Dickason, "Biographies in Naval History: Rear Admiral Grace Murray Hopper, USN, 9 December 1906–1 January 1992." Naval History and Heritage Command. http://www.history.navy.mil/bios/hopper_grace.htm, accessed August 20, 2014.

HERTHA AYRTON

210 *"astonished . . . one of their own sex"* Evelyn Sharp, *Hertha Ayrton: 1854–1923, a Memoir.* London: E. Arnold & Company, 1926.

210 *"Personally I do not agree"* Ibid.

211 *"for an original discovery"* "Hughes Medal." The Royal Society, https://royalsociety.org/awards/hughes-medal/, accessed August 17, 2014.

211 *"How can I answer"* Hertha Ayrton, Census Form for *Census of England and Wales, 1911*, in *Extraordinary Women in Science & Medicine: Four Centuries of Achievement.* An Exhibition at the Grolier Club, September 18–November 23, 2013.

211 *"An error that ascribes"* Evelyn Sharp, *Hertha Ayrton: 1854–1923, a Memoir.* London: E. Arnold & Company, 1926.

HEDY LAMARR

213 *"I've never been satisfied"* Gladys Hall, "The Life and Loves of Hedy Lamarr." Modern Romances, 1938. As cited in Richard Rhodes, *Hedy's Folly: The Life and Breakthrough Inventions of Hedy Lamarr, the Most Beautiful Woman in the World.* New York: Vintage Books, 2012.

213 *"[My father] made me"* Ibid.

214 *"Any girl can be glamorous"* Richard Schickel, *The Stars.* New York: Dial, 1962.

214 *"I don't like"* Hedy Lamarr, *Ecstasy and Me: My Life as a Woman.* New York: Fawcett Crest Book, 1967.

215 *"It was a flop"* Fleming Meeks, "I Guess They Just Take and Forget About a Person," *Forbes*, May 14, 1990. As cited in Richard Rhodes, *Hedy's Folly: The Life and Breakthrough Inventions of Hedy Lamarr, the Most Beautiful Woman in the World.* New York: Vintage Books, 2012.

217 *"Hedy Lamarr, screen actress"* "Hedy Lamarr Inventor." *New York Times*, October 1, 1941.

218 *"It's about time."* As quoted in Richard Rhodes, *Hedy's Folly: The Life and Breakthrough Inventions of Hedy Lamarr, the Most Beautiful Woman in the World.* New York: Vintage Books, 2012.

RUTH BENERITO

219 *"I'm not good with my hands"* Agricultural Research Service, US Department of Agriculture, "Conversations from the Hall of Fame." http://www.ars.usda.gov/is/video/asx/benerito.broadband.asx, accessed August 31, 2014.

220 *"Any number of people"* Ibid.

220 *"It was a good education"* Ibid.

220 *"I think that's what"* Ibid.

220 *"when [the government] put a lot of money"* Ibid.

220 *"I said I've been here"* Ibid.

STEPHANIE KWOLEK

222 *"pulling a string"* Paul W. Morgan and Stephanie L. Kwolek, "The Nylon Rope Trick: Demonstration of Condensation Polymerization." *Journal of Chemical Education*, April 1959.

224 *"It wasn't exactly"* Maureen Milford, "Mother of Invention Has Helped Save Thousands." *USA Today*, July 4, 2007.

224 *"I never in a thousand"* Ibid.

BIBLIOGRAPHY

INTRODUCTION

Charlotte to Lego Company, January 25, 2014, in Sociological Images, http://thesocietypages.org/socimages/2014/01/31/this-month-in-socimages-january-2014/.

Martin, Douglas, "Yvonne Brill, a Pioneering Rocket Scientist, Dies at 88." *New York Times*, March 30, 2013.

Sharp, Evelyn, *Hertha Ayrton: 1854–1923, a Memoir.* London: E. Arnold & Company, 1926.

Sherr, Lynn, *Sally Ride: America's First Woman in Space.* New York: Simon & Schuster, 2014.

MARY PUTNAM JACOBI

Bittel, Carla Jean, *Mary Putnam Jacobi and the Politics of Medicine in Nineteenth-Century America.* Chapel Hill: University of North Carolina Press, 2009.

Clarke, Edward H., *Sex in Education or, A Fair Chance for Girls.* Boston: James R. Osgood and Company, 1875.

Jacobi, Mary Putnam, *Life and Letters of Mary Putnam Jacobi.* New York: G. P. Putnam's Sons, 1925.

———, *The Question of Rest for Women During Menstruation.* New York: G. P. Putnam's Sons, 1877.

Smeltzer, Ronald K., Robert J. Ruben, and Paulette Rose, *Extraordinary Women in Science & Medicine: Four Centuries of Achievement.* New York: Grolier Club, 2013.

ANNA WESSELS WILLIAMS

Barry, John M., *The Great Influenza: The Story of the Deadliest Pandemic in History.* New York: Penguin Books, 2005.

Emrich, John, "Anna Wessels Williams, M.D.: Infectious Disease Pioneer and Public Health Advocate." *AAI Newsletter*, March/April 2012.

Morantz-Sanchez, Regina Markell, *Sympathy & Science: Women Physicians in American Medicine.* Chapel Hill: University of North Carolina Press, 2000.

"94 Retired by City; 208 More Will Go." *New York Times*, March 24, 1934.

Ogilvie, Marilyn Bailey, and Joy Dorothy Harvey, *The Biographical Dictionary of Women in Science: Pioneering Lives from Ancient Times to the Mid-Twentieth Century.* New York: Routledge, 2000.

Yount, Lisa, *A to Z of Women in Science and Math.* New York: Facts on File, 2008.

ALICE BALL

Encyclopædia Britannica Online. s. v. "leprosy." http://www.britannica.com/EBchecked/topic/336868/leprosy, accessed October 14, 2014.

London, Jack, *The Cruise of the Snark.* New York: Macmillan, 1911.

Wermager, Paul, and Carl Heltzel, "Alice A. Augusta Ball: Young Chemist Gave Hope to Millions." *ChemMatters*, February 2007.

GERTY RADNITZ CORI

McGrayne, Sharon Bertsch, *Nobel Prize Women in Science: Their Lives, Struggles, and Momentous Discoveries.* 2nd ed. Washington, DC: National Academies Press, 2001.

Smeltzer, Ronald K., Robert J. Ruben, and Paulette Rose, *Extraordinary Women in Science & Medicine: Four Centuries of Achievement.* New York: Grolier Club, 2013.

HELEN TAUSSIG

Altman, Lawrence K., "Dr. Helen Taussig, 87, Dies; Led in Blue Baby Operation." *New York Times*, May 22, 1986.

Bart, Jody, *Women Succeeding in the Sciences: Theories and Practices Across Disciplines.* West Lafayette, IN: Purdue Research Foundation, 2000.

Smeltzer, Ronald K., Robert J. Ruben, and Paulette Rose, *Extraordinary Women in Science & Medicine: Four Centuries of Achievement.* New York: Grolier Club, 2013.

Stevenson, Jeanne Hackley, "Helen Brooke Taussig, 1898: The 'Blue Baby' Doctor." *Notable Maryland Women.* Cambridge, MD: Tidewater, 1977.

ELSIE WIDDOWSON

Ashwell, Margaret, "Elsie May Widdowson, C.H., 21 October 1906–14 June 2000." Biographical Memoirs of Fellows of the Royal Society, December 1, 2002.

———, "Obituary: Elsie Widdowson (1906–2000)." *Nature*, August 24, 2000.

"Dr. Elsie Widdowson." *MRC Human Nutrition Research*, Elsie Widdowson Laboratory. http://www.mrc-hnr.cam.ac.uk/about-us/history/dr-elsie-widdowson-ch-cbe-frs/, accessed September 24, 2014.

Elliott, Jane, "Elsie—Mother of the Modern Loaf." *BBC News*, March 25, 2007.

"Elsie Widdowson." *Economist*, June 29, 2000.

"Elsie Widdowson." *Telegraph*, June 22, 2000.

Weaver, L. T., "Autumn Books: McCance and Widdowson—A Scientific Partnership of 60 Years." *Archives of Disease in Childhood*, 1993.

VIRGINIA APGAR

Apgar, Virginia, Letter to Allen O. Whipple. Mount Holyoke College, Archives and Special Collections, Virginia Apgar Papers, MS 0504, November 29, 1937.

Nee, Joseph F., "Eulogy—Memorial Service for Dr. Virginia Apgar." Mount Holyoke College. Archives and Special Collections. L. Stanley James Papers. MS 0782, Box 2, Folder 2: Correspondence about Apgar 1973–1975, September 15, 1974.

"The Virginia Apgar Papers: Biographical Information." U.S. National Library of Medicine. http://profiles.nlm.nih.gov/ps/retrieve/Narrative/CP/p-nid/178, accessed June 13, 2014.

"The Virginia Apgar Papers: Obstetric Anesthesia and a Scorecard for Newborns, 1949–1958." U.S. National Library of Medicine. http://profiles.nlm.nih.gov/ps/retrieve/Narrative/CP/p-nid/178, accessed June 13, 2014.

"The Virginia Apgar Papers: The National Foundation–March of Dimes, 1959–1974." US National Library of Medicine. http://profiles.nlm.nih.gov/ps/retrieve/Narrative/CP/p-nid/178, accessed June 13, 2014.

DOROTHY CROWFOOT HODGKIN

"Dorothy Crowfoot Hodgkin—Biographical." http://www.nobelprize.org/nobel_prizes/chemistry/laureates/1964/hodgkin-bio.html, accessed October 13, 2014.

McGrayne, Sharon Bertsch, *Nobel Prize Women in Science: Their Lives, Struggles, and Momentous Discoveries.* 2nd ed. Washington, DC: National Academies Press, 2001.

"The Nobel Prize in Chemistry 1964," Nobel Prize. http://www.nobelprize.org/nobel_prizes/chemistry/laureates/1964/, accessed August 15, 2014.

GERTRUDE BELLE ELION

"Gertrude B. Elion—Biographical." http://www.nobelprize.org/nobel_prizes/medicine/laureates/1988/elion-bio.html, accessed October 14, 2014.

McGrayne, Sharon Bertsch, *Nobel Prize Women in Science: Their Lives, Struggles, and Momentous Discoveries.* 2nd ed. Washington, DC: National Academies Press, 2001.

Smeltzer, Ronald K., Robert J. Ruben, and Paulette Rose, *Extraordinary Women in Science & Medicine: Four Centuries of Achievement.* New York: Grolier Club, 2013.

JANE WRIGHT

Chung, King-Thom, *Women Pioneers of Medical Research: Biographies of 25 Outstanding Scientists.* Jefferson, NC: McFarland, 2010.

"Homecoming for Jane Wright." *Ebony,* May 1968.

Jones, Alison, personal interview, September 14, 2014.

Piana, Ronald, "Jane Cooke Wright, MD, ASCO Cofounder, Dies at 93." *ASCO Post,* March 15, 2013.

Swain, Sandra M., "A Passion for Solving the Puzzle of Cancer: Jane Cooke Wright, M.D., 1919–2013." *Oncologist,* June 2013.

Warren, Wini, *Black Women Scientists in the United States.* Bloomington: Indiana University Press, 1999.

Webber, Bruce, "Jane Wright, Oncology Pioneer, Dies at 93." *New York Times,* March 2, 2013.

Wright, Jane C., "Cancer Chemotherapy: Past, Present, and Future—Part I." *JAMA,* August 1984.

Yount, Lisa, *Black Scientists.* New York: Facts on File, 1991.

MARIA SIBYLLA MERIAN

Haines, Catharine M. C., *International Women in Science: A Biographical Dictionary to 1950.* Santa Barbara, CA: ABC-CLIO, 2001.

"Maria Sibylla Merian: 1647–1717." National Museum of Women in the Arts. http://nmwa.org/explore/artist-profiles/maria-sibylla-merian, accessed September 7, 2014.

Todd, Kim, *Chrysalis: Maria Sibylla Merian and the Secrets of Metamorphosis.* Orlando, FL: Harcourt, 2007.

JEANNE VILLEPREUX-POWER

Brunner, Bernd, *The Ocean at Home: An Illustrated History of the Aquarium.* London: Reakton Books, 2003.

Encyclopædia Britannica Online. s. v. "Jeanne Villepreux-Power." http://www.britannica.com/EBchecked/topic/1759584/Jeanne-Villepreux-Power, accessed October 14, 2014.

Gage, Joslyn Matilda, "Woman as an Inventor." In *North American Review,* edited by Allen Thorndike Rice. New York: AMS Press, 1883.

Groeben, Christiane, "Tourists in Science: 19th Century Research Trips to the Mediterranean." *Proceedings of the California Academy of Sciences,* 2008.

Power, Jeannette, "Observations on the Habits of Various Marine Animals." *Annals and Magazine of Natural History.* London: Taylor & Francis, 1857.

MARY ANNING

Dickens, Charles, "Mary Anning, the Fossil Finder." *All the Year Round,* July 22, 1865.

Emling, Shelley, *The Fossil Hunter: Dinosaurs, Evolution, and the Woman Whose Discoveries Changed the World,* New York: Palgrave Macmillan, 2009.

Ogilvie, Marilyn Bailey, *Women in Science: Antiquity Through the Nineteenth Century.* Cambridge, MA: MIT Press, 1993.

ELLEN SWALLOW RICHARDS

Clarke, Robert, *Ellen Swallow: The Woman Who Founded Ecology.* Chicago: Follett Publishing Company, 1973.

"Ellen Swallow Richards." MIT History. http://libraries.mit.edu/mithistory/community/notable-persons/ellen-swallow-richards/, accessed August 30, 2014.

Hunt, Caroline Louisa, *The Life of Ellen H. Richards.* Boston: Whitcomb & Barrows, 1912.

Ogilvie, Marilyn Bailey, *Women in Science: Antiquity Through the Nineteenth Century.* Cambridge, MA: MIT Press, 1993.

Talbot, H. P., "Ellen Swallow Richards: Biography." *Technology Review,* 1911.

ALICE HAMILTON

"Alice Hamilton." *Chemical Heritage Foundation.* http://www.chemheritage.org/discover/online-resources/chemistry-in-history/themes/public-and-environmental-health/public-health-and-safety/richards-e.aspx, accessed May 19, 2014.

Hamilton, Alice, *Exploring the Dangerous Trades.* Boston: Little, Brown, 1943.

ALICE EVANS

Colwell, Rita R., "Alice C. Evans: Breaking Barriers." *Yale Journal of Biology and Medicine,* 1999.

Evans, Alice C, *Memoirs.* Unpublished, 1963. In Alice Catherine Evans. Papers, #2552, Division of Rare and Manuscript Collections. Cornell University Library.

Oakes, Elizabeth H., *Encyclopedia of World Scientists.* New York: Infobase, 2007.

United States Livestock Sanitary Association, Proceedings, Annual Meeting of the United States Livestock Sanitary Association, vols. 25–30, 1922.

Yount, Lisa, *A to Z of Women in Science and Math.* New York: Facts on File, 2008.

TILLY EDINGER

Buchholtz, Emily A., and Ernst-August Seyfarth, "The Gospel of the Fossil Brain: Tilly Edinger and the Science of Paleoneurology." *Brain Research Bulletin,* 1999.

———, "The Study of 'Fossil Brains': Tilly Edinger (1897–1967) and the Beginnings of Paleoneurology." *BioScience,* 2001.

RACHEL CARSON

Carson, Rachel, *Lost Woods: The Discovered Writing of Rachel Carson.* Edited by Linda Lear. Boston: Beacon Press, 1998.

———, *Silent Spring.* New York: Houghton Mifflin, 1962.

Lear, Linda, *Rachel Carson: Witness for Nature.* New York: Henry Holt, 1997.

Lewis, Jack, "The Birth of EPA." *EPA Journal,* November 1985.

Mahoney, Linda, "Rachel Carson (1907-1964)." National Women's History Museum. http://www.nwhm.org/education-resources/biography/biographies/rachel-carson/, accessed June 13, 2014.

"Rachel Carson Dies of Cancer. 'Silent Spring' Author Was 56." *New York Times,* April 15, 1964. http://www.nytimes.com/learning/general/onthisday/bday/0527.html, accessed June 13, 2014.

Rothman, Joshua, "Rachel Carson's Natural Histories." *The New Yorker,* September 27, 2012. http://www.newyorker.com/books/page-turner/rachel-carsons-natural-histories, accessed June 13, 2014.

Sideris, Lisa H., and Kathleen Dean Moore, eds., *Rachel Carson: Legacy and Challenge.* Albany: State University of New York Press, 2008.

RUTH PATRICK

"Almost Had a War on Her Hands." *Sydney Morning Herald,* August 11, 1960.

Bauers, Sandy, "Ruth Patrick: 'Den Mother of Ecology.'" *Philadelphia Inquirer,* March 5, 2007.

Belardo, Carolyn, "Pioneering Ecologist Dr. Ruth Patrick Dies." Academy of Natural Sciences of Drexel University, accessed September 1, 2014.

Dicke, William, "Ruth Patrick, a Pioneer in Science and Pollution Control Efforts, Is Dead at 105." *New York Times,* September 24, 2013. http://www.nytimes.com/2013/09/24/us/ruth-patrick-a-pioneer-in-pollution-control-dies-at-105.html?pagewanted=all&_r=0, accessed September 1, 2014.

"Dr. Ruth Patrick." WHYY. http://www.whyy.org/tv12/RuthPatrick.html, accessed September 1, 2014.

"Lecture 2: Biodiversity—Tom Lovejoy—Los Angeles." Reith Lectures, *BBC.* http://www.bbc.co.uk/radio4/reith2000/lecture2.shtml, accessed September 2, 2014.

Patrick, Ruth, "Some Diatoms of Great Salt Lake." *Bulletin of the Torrey Botanical Club,* March 1936.

Roddy, Michael, "Pollution Fears Come to Lakes, Springs." Associated Press, January 8, 1984.

NETTIE STEVENS

Brush, Stephen G., "Nettie M. Stevens and the Discovery of Sex Determination by Chromosomes." *Isis,* June 1978.

Cross, Patricia C., and John P. Steward, "Nettie Maria Stevens: Turn-of-the-Century Stanford Alumna Paved Path for Women in Biology." *Stanford Historical Society*, Winter 1993.

Gilbert, S. F., *Developmental Biology*. 6th ed. Sunderland, MA: Sinauer Associates, 2000. Chapter 17, Sex Determination. Available from: http://www.ncbi.nlm.nih.gov/books/NBK9985/.

Hagen, Joel B., *Doing Biology*. New York: HarperCollins College Publishers, 1996.

Hake, Laura, "Genetic Mechanisms of Sex Determination." *Nature Education*, 2008.

"Nettie Stevens Uses Diptera to Describe Two Heterochromosomes." *An American Amalgam: The Chromosome Theory of Heredity*. Cold Spring Harbor, NY: Cold Spring Harbor Laboratory Press, 2004.

Ogilvie, Marilyn Bailey, *Women in Science: Antiquity Through the Nineteenth Century*. Cambridge, MA: MIT Press, 1993.

Stevens, N. M., *Studies in Spermatogenesis, Part I & Part II*. Washington DC: Carnegie Institution of Washington, 1905 & 1906.

HILDE MANGOLD

Doty, Maria, "Hilde Mangold (1898–1924)." Embryo Project at Arizona State University. http://embryo.asu.edu/pages/hilde-mangold-1898-1924, accessed September 20, 2014.

Hagan, Joel B., "Nettie Stevens & the Problem of Sex Determination." http://www1.umn.edu/ships/db/stevens.pdf, accessed September 22, 2014.

Hamburger, Viktor, "Hilde Mangold: Co-Discoverer of the Organizer." *Journal of the History of Biology*, Spring 1984.

Sander, Klaus, and Peter E. Faessler, "Introducing the Spemann-Mangold Organizer: Experiments and Insights That Generated a Key Concept in Developmental Biology." *International Journal of Developmental Biology*, 2001.

CHARLOTTE AUERBACH

Beale, G. H., "Charlotte Auerbach, 14 May 1899–17 March 1994." *Biographical Memoirs of Fellows of the Royal Society*, November 1995.

Haines, Catharine M. C., *International Women in Science: A Biographical Dictionary to 1950*. Santa Barbara, CA: ABC-CLIO, 2001.

BARBARA MCCLINTOCK

Keller, Evelyn Fox, *A Feeling for the Organism: The Life and Work of Barbara McClintock*. New York: Henry Holt, 1983.

McGrayne, Sharon Bertsch, *Nobel Prize Women in Science: Their Lives, Struggles, and Momentous Discoveries*. 2nd ed. Washington, DC: National Academies Press, 2001.

Nobel Prize, http://www.nobelprize.org/nobel_prizes/medicine/laureates/1983/, accessed August 4, 2014.

SALOME GLUECKSOHN WAELSCH

Ambrose, Susan A., et al., *Journeys of Women in Science and Engineering: No Universal Constants*. Philadelphia: Temple University Press, 1997.

Gluecksohn-Schoenheimer, S., "The Development of Two Tailless Mutants." *Genetics*, 1938.

Solter, Davor, "In Memoriam: Salome Gluecksohn-Waelsch (1907–2007)." *Developmental Cell*, January 2008.

Waelsch, Salome, "The Causal Analysis of Development in the Past Half Century: A Personal History." *Development*, 1992.

RITA LEVI-MONTALCINI

McGrayne, Sharon Bertsch, *Nobel Prize Women in Science: Their Lives, Struggles, and Momentous Discoveries*. 2nd ed. Washington, DC: National Academies Press, 2001.

Smeltzer, Ronald K., Robert J. Ruben, and Paulette Rose, *Extraordinary Women in Science & Medicine: Four Centuries of Achievement*. New York: Grolier Club, 2013.

ROSALIND FRANKLIN

Maddox, Brenda, *Rosalind Franklin: The Dark Lady of DNA*. New York: HarperCollins, 2002.

McGrayne, Sharon Bertsch, *Nobel Prize Women in Science: Their Lives, Struggles, and Momentous Discoveries*. 2nd ed. Washington, DC: National Academies Press, 2001.

Watson, James, *The Double Helix: A Personal Account of the Discovery of the Structure of DNA*. New York: Scribner, 1968.

ANNE MCLAREN

Biggers, John D., "Research in the Canine Block." *International Journal of Developmental Biology*, 2001.

Clarke, Ann G., "Anne McLaren—A Tribute from Her Research Students." *International Journal of Developmental Biology*, 2001.

Hogan, Brigid, "From Embryo to Ethics: A Career in Science and Social Responsibility: An Interview with Anne McLaren." *International Journal of Developmental Biology*, 2001.

Renfree, Marilyn, and Roger Short, "In Memoriam." *International Journal of Developmental Biology*, 2008.

Surani, Azim, and Jim Smith, "Anne McLaren (1927–2007)." *Nature*, August 16, 2007.

LYNN MARGULIS

Lake, James A., "Lynn Margulis (1938–2011)." *Nature*, December 22, 2011.

Margulis, Lynn, "Chapter 7: Lynn Margulis." In John Brockman [ed.], *The Third Culture: Beyond the Scientific Revolution*. New York: Simon & Schuster, 1995.

Rose, Steven, "Lynn Margulis," obituary. *Guardian*, December 11, 2011.

Rutgers Research Channel, "Lynn Margulis 2004 Rutgers Interview." https://www.youtube.com/watch?v=b8xqu_TlQPU>.

Sagan, Dorion, *Lynn Margulis: The Life and Legacy of a Scientific Rebel*. White River Junction, VT: Chelsea Green, 2012.

Teresi, Dick, "Discover Interview: Lynn Margulis Says She's Not Controversial, She's Right," *Discover*, June 17, 2011.

Webber, Bruce, "Lynn Margulis, Evolution Theorist, Dies at 73." *New York Times*, November 25, 2011. http://www.nytimes.com/2011/11/25/science/lynn-margulis-trailblazing-theorist-on-evolution-dies-at-73.html?_r=0&pagewanted=print.

ÉMILIE DU CHÂTELET

du Châtelet, Émilie, "Translator's Preface for *The Fable of the Bees*." In Judith P. Zinsser (ed.), *Émilie du Châtelet: Selected Philosophical and Scientific Writings*. Chicago: University of Chicago Press, 2009.

Ogilvie, Marilyn Bailey, *Women in Science: Antiquity Through the Nineteenth Century*. Cambridge, MA: MIT Press, 1993.

Smeltzer, Ronald K., Robert J. Ruben, and Paulette Rose, *Extraordinary Women in Science & Medicine: Four Centuries of Achievement.* New York: Grolier Club, 2013.

Zinsser, Judith P., *Émilie du Châtelet: Daring Genius of the Enlightenment.* New York: Penguin Books, 2007.

LISE MEITNER

McGrayne, Sharon Bertsch, *Nobel Prize Women in Science: Their Lives, Struggles, and Momentous Discoveries.* 2nd ed. Washington, DC: National Academies Press, 2001.

Meitner, Lise, "Looking Back." *Bulletin of the Atomic Scientists*, November 1964.

IRÈNE JOLIOT-CURIE

Interview with Lew Kowarski by Charles Weiner, Niels Bohr Library & Archives, American Institute of Physics, College Park, MD, www.aip.org/history/ohilist/4717_1.html.

McGrayne, Sharon Bertsch, *Nobel Prize Women in Science: Their Lives, Struggles, and Momentous Discoveries.* 2nd ed. Washington, DC: National Academies Press, 2001.

"Mlle. Curie Reads Thesis." *New York Times*, March 31, 1925.

MARIA GOEPPERT MAYER

Born, Max. *My Life: Recollections of a Nobel Laureate*, New York: Scribner, 1978.

Mayer, Maria Goeppert, "The Shell Model." Nobel Lecture, December 12, 1963.

McGrayne, Sharon Bertsch, *Nobel Prize Women in Science: Their Lives, Struggles, and Momentous Discoveries.* 2nd ed. Washington, DC: National Academies Press, 2001.

Sachs, Robert G., "Maria Goeppert Mayer." In Edward Shils [ed.], *Remembering the University of Chicago: Teachers, Scientists, and Scholars.* Chicago: University of Chicago Press, 1991.

Smeltzer, Ronald K., Robert J. Ruben, and Paulette Rose, *Extraordinary Women in Science & Medicine: Four Centuries of Achievement.* New York: Grolier Club, 2013.

MARGUERITE PEREY

Byers, Nina, and Gary Williams, *Out of the Shadows: Contributions of Twentieth-Century Women to Physics*. New York: Cambridge University Press, 2006.

Haines, Catharine M. C., *International Women in Science: A Biographical Dictionary to 1950*. Santa Barbara, CA: ABC-CLIO, 2001.

"Madame Curie's Assistant: Scientific Battle Won, She's Losing Medical One." *Milwaukee Journal*, July 15, 1962.

Rayner-Canham, Marelene F., and Geoffrey Rayner-Canham, *Women in Chemistry: Their Changing Roles from Alchemical Times to the Mid-Twentieth Century*. Philadelphia: Chemical Heritage Foundation, 2001.

CHIEN-SHIUNG WU

McGrayne, Sharon Bertsch, *Nobel Prize Women in Science: Their Lives, Struggles, and Momentous Discoveries*. 2nd ed. Washington, DC: National Academies Press, 2001.

Smeltzer, Ronald K., Robert J. Ruben, and Paulette Rose, *Extraordinary Women in Science & Medicine: Four Centuries of Achievement*. New York: Grolier Club, 2013.

ROSALYN SUSSMAN YALOW

McGrayne, Sharon Bertsch, *Nobel Prize Women in Science: Their Lives, Struggles, and Momentous Discoveries*. 2nd ed. Washington, DC: National Academies Press, 2001.

Smeltzer, Ronald K., Robert J. Ruben, and Paulette Rose, *Extraordinary Women in Science & Medicine: Four Centuries of Achievement*. New York: Grolier Club, 2013.

MARIA MITCHELL

Mitchell, Maria, *Maria Mitchell: Life, Letters, and Journals*. Boston: Lee & Shepard, 1896.

"This Month in Physics History, Maria Mitchell Discovers a Comet." *American Physical Society*. http://www.aps.org/publications/apsnews/200610/history.cfm, accessed November 7, 2013.

Vassar Historian, "Vassar Encyclopedia: Maria Mitchell." *The Vassar Encyclopedia*. http://vcencyclopedia.vassar.edu/faculty/original-faculty/maria-mitchell1.html, accessed November 7, 2013.

ANNIE JUMP CANNON

Bok, Priscilla F., "Annie Jump Cannon, 1863–1941." *Publications of the Astronomical Society of the Pacific*, June 1941.

Bruck, H. A., "Obituary: Dr. Annie J. Cannon." *Observatory*, 1941.

"Delaware Daughter Star Gazer." *Delmarva Star*, March 11, 1934.

"Dr. Annie Cannon Called 'One of 12 Greatest Living Women.'" *Milwaukee Journal*, April 7, 1936.

Ogilvie, Marilyn Bailey, *Women in Science: Antiquity Through the Nineteenth Century*. Cambridge, MA: MIT Press, 1993.

INGE LEHMANN

Bolt, Bruce A., "50 Years of Studies on the Inner Core." *History of Geophysics*, February 10, 1987.

———, "Inge Lehmann, Contributions of Women to Physics." http://www.physics.ucla.edu/~cwp/articles/bolt.html, accessed September 11, 2014.

Interview of Jack Oliver by Ron Doel on September 27, 1997. Niels Bohr Library & Archives, American Institute of Physics, College Park, MD, http://www.aip.org/history/ohilist/6928_2.html.

Mathez, Edmond A., *Earth: Inside and Out*. New York: New Press, 2000.

Ogilvie, Marilyn Bailey, and Joy Dorothy Harvey, *The Biographical Dictionary of Women in Science: Pioneering Lives from Ancient Times to the Mid-Twentieth Century*. New York: Routledge, 2000.

Rousseau, Christiane, "How Inge Lehmann Discovered the Inner Core of the Earth." *College Mathematics Journal*, November 2013.

MARIE THARP

Felt, Hali, "Marie Tharp: Portrait of a Scientist." *General Bathymetric Chart of the Oceans*. http://www.gebco.net/about_us/gebco_science_day/documents/gebco_sixth_science_day_felt.pdf, accessed September 10, 2014.

———, *Soundings: The Story of the Remarkable Woman Who Mapped the Ocean Floor*. New York: Henry Holt, 2012.

Fox, Margalit, "Marie Tharp, Oceanographic Cartographer, Dies at 86." *New York Times*, August 26, 2006.

Hall, Stephen S., "The Contrary Map Maker." *New York Times*, December 31, 2006.

Levin, Tanya, "Oral History Transcript—Dr. Marie Tharp." *American Institute of Physics*. http://www.aip.org/history/ohilist/22896_4.html, accessed September 10, 2014.

"Ocean Explorer: Soundings, Sea-Bottom, and Geophysics." National Oceanic and Atmospheric Administration. http://oceanexplorer.noaa.gov/history/quotes/soundings/soundings.html, accessed September 10, 2014.

"Remembered: Marie Tharp, Pioneering Mapmaker of the Ocean Floor." Earth Institute at Columbia University. http://www.earth.columbia.edu/news/2006/story08-24-06.php, accessed September 10, 2014.

Tharp, Marie, "Connect the Dots: Mapping the Seafloor and Discovering the Mid-Ocean Ridge." In *Lamont-Doherty Earth Observatory of Columbia: Twelve Perspectives on the First Fifty Years 1949–1999*, edited by Laurence Lippsett. Palisades, NY: Lamont-Doherty Earth Observatory of Columbia University, 1999.

YVONNE BRILL

Martin, Douglas, "Yvonne Brill, a Pioneering Rocket Scientist, Dies at 88." *New York Times*, March 30, 2013.

Rice, Deborah, "Interview with Yvonne Brill on November 3rd, 2005." Society of Women Engineers. http://www.djgcreate.com/swe/joomla/images/stories/brill/BRILLBRILL.pdf, accessed October 26, 2013.

Wayne, Tiffany K., *American Women of Science Since 1900*. Santa Barbara, CA: ABC-CLIO, 2011.

SALLY RIDE

"An Interview with Sally Ride." *Nova* PBS. https://www.youtube.com/watch?v=yb6vw9AmiLs, accessed August 30, 2014.

Grady, Denise, "American Woman Who Shattered Space Ceiling." *New York Times*, July 23, 2012.

Knipfer, Cody, "Sally Ride and Valentina Tereshkova: Changing the Course of Human Space Exploration." NASA. http://www.nasa.gov/topics/history/features/ride_anniversary.html#.VDwXddR4pfF, accessed August 30, 2014.

"Mission to Planet Earth." NASA. http://www.hq.nasa.gov/office/nsp/mtpe.htm, accessed August 30, 2014

Ride, Sally, *NASA: Leadership and America's Future in Space*, August 1987.

Sherr, Lynn, *Sally Ride: America's First Woman in Space*. New York: Simon & Schuster, 2014.

MARIA GAETANA AGNESI

Alexanderson, Gerald L., "About the Cover: Maria Gaetana Agnesi—A Divided Life." *Bulletin of the American Mathematical Society*, January 2013.

Mazzotti, Massimo, *The World of Maria Gaetana Agnesi, Mathematician of God*. Baltimore,MD: Johns Hopkins University Press, 2007.

Ogilvie, Marilyn Bailey, *Women in Science: Antiquity Through the Nineteenth Century*. Cambridge, MA: MIT Press, 1993.

Smeltzer, Ronald K., Robert J. Ruben, and Paulette Rose, *Extraordinary Women in Science & Medicine: Four Centuries of Achievement*. New York: Grolier Club, 2013.

Stigler, Stephen M., *Statistics on the Table: The History of Statistical Concepts and Methods*. Cambridge, MA: Harvard University Press, 1999.

ADA LOVELACE

"Charles Babbage: Pioneer of the Digital Age: An Exhibition at the Beinecke Library." Yale University Beinecke Rare Books & Manuscript Library. http://beinecke.library.yale.edu/exhibitions/charles-babbage-pioneer-digital-age-exhibition-beinecke-library, accessed September 15, 2014.

Charman-Anderson, Suw, "Ada Lovelace: Victorian Computing Visionary." *Finding Ada*. http://findingada.com/book/ada-lovelace-victorian-computing-visionary/, accessed August 29, 2014

Menabrea, L. F., "Sketch of the Analytical Engine Invented by Charles Babbage, Esq.," trans. Augusta Ada Byron King, Countess of Lovelace, *Scientific Memoirs*, 1843.

Morais, Betsy, "Ada Lovelace: The First Tech Visionary." *The New Yorker*, October 15, 2013. http://www.newyorker.com/tech/elements/ada-lovelace-the-first-tech-visionary, accessed August 28, 2014.

Smeltzer, Ronald K., Robert J. Ruben, and Paulette Rose, *Extraordinary Women in Science & Medicine: Four Centuries of Achievement*. New York: Grolier Club, 2013.

Stein, Dorothy, *Ada: A Life and a Legacy*. Cambridge, MA: MIT Press, 1985.

FLORENCE NIGHTINGALE

Bostridge, Mark, *Florence Nightingale: The Making of an Icon*. New York: Farrar, Straus & Giroux, 2008.

Nelson, Sioban, and Anne Marie Fafferty, *Notes on Nightingale*. Ithaca, NY: ILR Press, 2010.

Nightingale, Florence, *Collected Works of Florence Nightingale.* Waterloo, Ontario, Canada: Wilfrid Laurier University Press, 2009.

Smeltzer, Ronald K., Robert J. Ruben, and Paulette Rose, *Extraordinary Women in Science & Medicine: Four Centuries of Achievement.* New York: Grolier Club, 2013.

SOPHIE KOWALEVSKI

Cooke, Roger, *The Mathematics of Sonya Kovalevskaya.* New York: Springer-Verlag, 1984.

Cooke, Roger L., "The Life of S. V. Kovalevskaya." In Vadim Kuznetsov [ed.], *The Kowalevski Property.* Providence, RI: American Mathematical Society, 2002.

Kovalevskaya, Sofya, *A Russian Childhood.* Translated by Beatrice Stillman, assisted by P. Y. Kochina. New York: Springer, 1978.

Ogilvie, Marilyn Bailey, *Women in Science: Antiquity Through the Nineteenth Century.* Cambridge, MA: MIT Press, 1993.

Smeltzer, Ronald K., Robert J. Ruben, and Paulette Rose, *Extraordinary Women in Science & Medicine: Four Centuries of Achievement.* New York: Grolier Club, 2013.

EMMY NOETHER

Angier, Natalie, "The Mighty Mathematician You've Never Heard Of." *New York Times,* March 26, 2012.

Byers, Nina, and Gary Williams, *Out of the Shadows: Contributions of Twentieth-Century Women to Physics.* New York: Cambridge University Press, 2006.

Einstein, Albert, "The Late Emmy Noether." *New York Times,* May 4, 1935.

McGrayne, Sharon Bertsch, *Nobel Prize Women in Science: Their Lives, Struggles, and Momentous Discoveries.* 2nd ed. Washington, DC: National Academies Press, 2001.

Smeltzer, Ronald K., Robert J. Ruben, and Paulette Rose, *Extraordinary Women in Science & Medicine: Four Centuries of Achievement.* New York: Grolier Club, 2013.

MARY CARTWRIGHT

Albers, Donald J., and Gerald L. Alexanderson, *Fascinating Mathematical People: Interviews and Memoirs.* Princeton, NJ: Princeton University Press, 2011.

Davis, Philip J., "Snapshots of a Lively Character: Mary Lucy Cartwright, 1900–1998." *Society for Industrial and Applied Mathematics*. http://www.siam.org/news/news.php?id=863, accessed September 12, 2014.

Dyson, Freeman J., "Mary Lucy Cartwright." In Nina Byers and Gary Williams [eds.], *Out of the Shadows: Contributions of Twentieth-Century Women to Physics*. New York: Cambridge University Press, 2006.

Haines, Catharine M. C., *International Women in Science: A Biographical Dictionary to 1950*. Santa Barbara, CA: ABC-CLIO, 2001.

Jardine, Lisa, "A Point of View: Mary, Queen of Maths." *BBC*, March 8, 2013.

McMurran, Shawnee, and James Tattersall, "Mary Cartwright (1900–1998)." *Notices of the AMS*, February 1999.

Rees, Joan, "Obituary: Dame Mary Cartwright." *Independent*, April 9, 1998. http://www.independent.co.uk/news/obituaries/obituary-dame-mary-cartwright-1155320.html.

GRACE MURRAY HOPPER

Ceruzzi, Paul, Introduction to *A Manual of Operation for the Automatic Sequence Controlled Calculator*. Cambridge, MA: MIT Press, 1946.

Hamblen, Diane, Grace M. Hopper, and Elizabeth Dickason, "Biographies in Naval History: Rear Admiral Grace Murray Hopper, USN, 9 December 1906–1 January 1992." Naval History and Heritage Command. http://www.history.navy.mil/bios/hopper_grace.htm, accessed August 20, 2014.

Merzbach, Uta C., "Computer Oral History Collection, Grace Murray Hopper (1906–1992)." Computer Oral History Collection, 1969–1973, 1977, Archives Center, National Museum of American History, July 1968.

Smeltzer, Ronald K., Robert J. Ruben, and Paulette Rose, *Extraordinary Women in Science & Medicine: Four Centuries of Achievement*. New York: Grolier Club, 2013.

HERTHA AYRTON

Ayrton, Hertha, Census Form for *Census of England and Wales, 1911*, in *Extraordinary Women in Science & Medicine: Four Centuries of Achievement*. An Exhibition at the Grolier Club, September 18–November 23, 2013.

Byers, Nina, and Gary Williams, *Out of the Shadows: Contributions of Twentieth-Century Women to Physics*. New York: Cambridge University Press, 2006.

Grinstein, Louise S., Rose K. Rose, and Miriam H. Rafailovich, eds., *Women in Chemistry and Physics*. Westport, CT: Greenwood Press, 1993.

"Hughes Medal." The Royal Society. https://royalsociety.org/awards/hughes-medal/ accessed August 17, 2014.

Ogilvie, Marilyn Bailey, *Women in Science: Antiquity Through the Nineteenth Century.* Cambridge, MA: MIT Press, 1993.

Sharp, Evelyn, *Hertha Ayrton: 1854–1923, a Memoir.* London: E. Arnold & Company, 1926.

Smeltzer, Ronald K., Robert J. Ruben, and Paulette Rose, *Extraordinary Women in Science & Medicine: Four Centuries of Achievement.* New York: Grolier Club, 2013.

HEDY LAMARR

George, Antheil, and Markey Hedy Kiesler, assignee, Secret Communication System, Patent 2292387 A. August 11, 1942.

"Hedy Lamarr Inventor." *New York Times*, October 1, 1941.

Rhodes, Richard, *Hedy's Folly: The Life and Breakthrough Inventions of Hedy Lamarr, the Most Beautiful Woman in the World.* New York: Vintage Books, 2012.

RUTH BENERITO

Agricultural Research Service, US Department of Agriculture, "Conversations from the Hall of Fame." http://www.ars.usda.gov/is/video/asx/benerito.broadband.asx, accessed August 31, 2014.

Condon, Brian D., and J. Vincent Edwards, "Cross-Linking Cotton." *Agricultural Research*, February 2009.

Fox, Margalit, "Ruth Benerito, Who Made Cotton Cloth Behave, Dies at 97." *New York Times*, October 7, 2013.

"Ruth Benerito." Lemelson-MIT Program, Massachusetts Institute of Technology, Cambridge, MA, August 31, 2014. http://lemelson.mit.edu/winners/ruth-benerito.

Wolf, Lauren K., "Wrinkle-Free Cotton." *Chemical & Engineering News*, American Chemical Society, December 2, 2013.

Yafa, Stephen, *Cotton: The Biography of a Revolutionary Fiber.* New York: Penguin Group, 2005.

STEPHANIE KWOLEK

Lemelson Foundation, "1999 Lemelson-MIT Lifetime Achievement Award Winner Stephanie L. Kwolek." http://youtube/8dX3Z5CyF3c, accessed February 28, 2009.

Milford, Maureen, "Mother of Invention Has Helped Save Thousands." *USA Today*, July 4, 2007.

Morgan, Paul W., and Stephanie L. Kwolek, "The Nylon Rope Trick: Demonstration of Condensation Polymerization." *Journal of Chemical Education*, April 1959.

Norton, Tucker, personal interview, February 16, 2011.

Pearce, Jeremy, "Stephanie L. Kwolek, Inventor of Kevlar, Is Dead at 90." *New York Times*, June 20, 2014.

"Stephanie L. Kwolek." Chemical Heritage Foundation. http://www.chemheritage.org/discover/online-resources/chemistry-in-history/themes/petrochemistry-and-synthetic-polymers/synthetic-polymers/kwolek.aspx, accessed August 27, 2014.

INDEX

CREDITS

Grateful acknowledgment is made to the following:

Ernst Mayr Library, Museum of Comparative Zoology Archives, Harvard University: poem written in the letter from Glen Jepsen to Tilly Edinger, June 11, 1957. Courtesy of the Ernst Mayr Library, Museum of Comparative Zoology Archives, Harvard University.

March of Dimes Foundation: excerpt from Joseph F. Nee's memorial service speech for Virginia Apgar, September 15, 1974, from The Virginia Apgar Papers, U.S. National Library of Medicine, New York. All rights reserved. Reprinted by permission of the March of Dimes.

Philadelphia Inquirer: excerpt from *Ruth Patrick: "The Den Mother of Ecology"* by Sandy Bauers, March 5, 2007, copyright © 2015. All rights reserved. Reprinted by permission of the *Philadelphia Inquirer.*

Rutgers, the State University of New Jersey: excerpt from the video *Lynn Margulis 2004 Rutgers Interview.* https://www.youtube.com/watch?v=b8xqu_TIQPU. All rights reserved. Reprinted by permission of Rutgers, the State University of New Jersey.

Springer: excerpt from "Hilde Mangold, Co-Discoverer of the Organizer" by Viktor Hamburger, from *The Journal of the History of Biology*, Vol. 17, Issue 1 (January 1, 1984). All rights reserved. Reprinted by permission of Springer.

The American Institute of Physics: excerpt from Dr. Lew Kowarski's oral history transcript, from the interview of Dr. Lew Kowarski by Dr. Charles Weiner, March 20, 1969, and excerpt from Dr. Jack Oliver's oral history regarding Inge Lehmann, from the interview of Dr. Jack Oliver by Ron Doel, September 27, 1997. All rights reserved. Reprinted by permission of the Niels Bohr Library & Archives, American Institute of Physics, College Park, MD, USA, www.aip.org/history/ohilist/LINK.

The British Library Board: excerpt from *Stein, Dorothy, Ada: A Life and Legacy* from MIT Press, 1985, MS 37182-37201. All rights reserved. Reprinted by permission of the British Library Board.

The Carnegie Institution for Science: excerpt from Thomas Hunt Morgan's letter of recommendation included as a part of Nettie M. Stevens' application for a Carnegie research grant, from the archives database of the Carnegie Institution for Science. Box number 24, folder number 30. June 27, 1903–June 1978. All rights reserved. Reprinted by permission of the Carnegie Institution for Science.

The Lilly Library: excerpt from a Western Union Cable from H. J. Muller to Charlotte Auerbach, June 21, 1941. Boxes 16–30, Auerbach, Charlotte, 1938–1967, Muller mss. Courtesy of The Lilly Library, Indiana University, Bloomington, Indiana.

The Milwaukee Journal: excerpt from *Madame Curie's Assistant: Scientific Battle Won; She's Losing Medical One* by Marguerite Perey, July 15, 1962. All rights reserved. Reprinted by permission of *The Milwaukee Journal.*

The National Academies Press: excerpt from *Nobel Prize Women in Science: Their Lives, Struggles, and Momentous Discoveries* by

Sharon Bertsch McGrayne, copyright © 2001 by the National Academies of Sciences. All rights reserved. Reprinted by permission of the National Academies Press.

The Royal Society: excerpt from "Charlotte Auerbach: 14 May 1899–17 March 1994" by G. H. Beale, from *Biographical Memoirs of Fellows of the Royal Society*, Vol. 41 (November 1995), pgs. 20–42. All rights reserved. Reprinted by permission of the Royal Society.

This I Believe, Inc.: copyright © 1952 by Gerty Cori, part of the This I Believe Essay Collection found at www.thisibelieve.org, copyright © 2005–2015, This I Believe, Inc. Used with permission.

Wayne State University: excerpt from the *Yvonne Brill Oral History Transcript*, from Engineering Pioneers Oral History: Yvonne Brill, Profiles of SWE Pioneers Oral History Project, Walter P. Reuther Library and Archives of Labor and Urban Affairs. All rights reserved. Reprinted by permission of the Walter P. Reuther Library and Archives of Labor and Urban Affairs, Wayne State University.